SpringerBriefs in Electrical and Computer Engineering

Speech Technology

Series editor

Amy Neustein, Fort Lee, NJ, USA

Editor's Note

The authors of this series have been hand-selected. They comprise some of the most outstanding scientists—drawn from academia and private industry—whose research is marked by its novelty, applicability, and practicality in providing broad based speech solutions. The SpringerBriefs in Speech Technology series provides the latest findings in speech technology gleaned from comprehensive literature reviews and *empirical investigations* that are performed in both laboratory and *real life* settings. Some of the topics covered in this series include the presentation of real life commercial deployment of spoken dialog systems, contemporary methods of speech parameterization, developments in information security for automated speech, forensic speaker recognition, use of sophisticated speech analytics in call centers, and an exploration of new methods of soft computing for improving human-computer interaction. Those in academia, the private sector, the self service industry, law enforcement, and government intelligence, are among the principal audience for this series, which is designed to serve as an important and essential reference guide for speech developers, system designers, speech engineers, linguists and others. In particular, a major audience of readers will consist of researchers and technical experts in the automated call center industry where speech processing is a key component to the functioning of customer care contact centers.

Amy Neustein, Ph.D., serves as Editor-in-Chief of the International Journal of Speech Technology (Springer). She edited the recently published book "Advances in Speech Recognition: Mobile Environments, Call Centers and Clinics" (Springer 2010), and serves as quest columnist on speech processing for Womensenews. Dr. Neustein is Founder and CEO of Linguistic Technology Systems, a NJ-based think tank for intelligent design of advanced natural language based emotion-detection software to improve human response in monitoring recorded conversations of terror suspects and helpline calls. Dr. Neustein's work appears in the peer review literature and in industry and mass media publications. Her academic books, which cover a range of political, social and legal topics, have been cited in the Chronicles of Higher Education, and have won her a pro Humanitate Literary Award. She serves on the visiting faculty of the National Judicial College and as a plenary speaker at conferences in artificial intelligence and computing. Dr. Neustein is a member of MIR (machine intelligence research) Labs, which does advanced work in computer technology to assist underdeveloped countries in improving their ability to cope with famine, disease/illness, and political and social affliction. She is a founding member of the New York City Speech Processing Consortium, a newly formed group of NY-based companies, publishing houses, and researchers dedicated to advancing speech technology research and development.

More information about this series at http://www.springer.com/series/10043

Nilanjan Dey · Amira S. Ashour

Direction of Arrival Estimation and Localization of Multi-Speech Sources

 Springer

Nilanjan Dey
Department of Information Technology
Techno India College of Technology
Kolkata
India

Amira S. Ashour
Department of Electronics and Electrical
   Communication Engineering
Faculty of Engineering
Tanta University
Tanta
Egypt

ISSN 2191-8112   ISSN 2191-8120   (electronic)
SpringerBriefs in Electrical and Computer Engineering
ISSN 2191-737X   ISSN 2191-7388   (electronic)
SpringerBriefs in Speech Technology
ISBN 978-3-319-73058-5   ISBN 978-3-319-73059-2   (eBook)
https://doi.org/10.1007/978-3-319-73059-2

Library of Congress Control Number: 2017961747

© The Author(s) 2018
This work is subject to copyright. All rights are reserved by the Publisher, whether the whole or part
of the material is concerned, specifically the rights of translation, reprinting, reuse of illustrations,
recitation, broadcasting, reproduction on microfilms or in any other physical way, and transmission
or information storage and retrieval, electronic adaptation, computer software, or by similar or dissimilar
methodology now known or hereafter developed.
The use of general descriptive names, registered names, trademarks, service marks, etc. in this
publication does not imply, even in the absence of a specific statement, that such names are exempt from
the relevant protective laws and regulations and therefore free for general use.
The publisher, the authors and the editors are safe to assume that the advice and information in this
book are believed to be true and accurate at the date of publication. Neither the publisher nor the
authors or the editors give a warranty, express or implied, with respect to the material contained herein or
for any errors or omissions that may have been made. The publisher remains neutral with regard to
jurisdictional claims in published maps and institutional affiliations.

Printed on acid-free paper

This Springer imprint is published by Springer Nature
The registered company is Springer International Publishing AG
The registered company address is: Gewerbestrasse 11, 6330 Cham, Switzerland

Preface

Speech processing and localization/tracking of acoustic sources have a significant role in the automation of several applications, including video conferencing with audio-based camera steering systems as well as surveillance systems. In such applications, it is essential to localize the speaker as well as any acoustic experience. Furthermore, localizing noise sources around/in a moving car environment is an active research area. These applications require preprocessing stage for speech enhancement based on automatic Direction of Arrival estimation (DOAE) of speech sources. Multi-DOAE is indispensable in real acoustic environments, such as mobile active speech sources.

Several outstanding DOAE techniques, such as Maximum Likelihood (ML) method, estimation of signal parameters via invariance techniques (ESPRIT), multiple signal classification (MUSIC), and Local Polynomial Approximation (LPA), can be employed in the speech sources DOAE and localization. Currently, the DOAE and localization contexts have an outstanding theoretical basis for several practical applications; however, it is still an embryonic research domain.

This book supports the researchers, designers, and engineers in various interdisciplinary domains, such as engineering, speech processing, mobile communication, direction of arrival estimation, and localization to explore the broad vision of the DOAE/localization of speech sources. The book introduces the concept and model of the acoustic sources. Then, it highlights the most contemporary studies on this pervasive problem. The book provides a brief overview of the most classical direction of arrival estimation and localization techniques. In addition, employing the optimization algorithms to improve the DOAE techniques is also highlighted. The book addressed the concept and principles of the multi-DOAE approaches. Using a microphone array, this book introduced the localization and tracking problem of multiple speech/acoustic sources. It includes applications of speech sources localization based on the DOAE approaches. The book reports the challenges facing the DOAE techniques in speech sources localization.

The unique features of this book include:

- Provides a solid background on the concept and model of the acoustical signal and sources.
- Offers a brief overview of the most classical direction of arrival estimation and localization techniques.
- Explores the role of optimization algorithms to improve the DOAE techniques.
- Highlights the concept and principles of the multi-DOAE approaches.
- Introduces the localization and tracking problem of multiple speech/acoustic sources with highlighting the most contemporary studies on this pervasive problem.
- Discusses several applications and real-life speech sources localization based on the DOAE approaches.
- Reports the challenges facing the DOAE techniques in speech sources localization.

Kolkata, India   Nilanjan Dey Ph.D.
Tanta, Egypt   Amira S. Ashour Ph.D.

Acknowledgements

Effective algorithms make assumptions, show a bias toward a simple solutions, trade off the costs of error against the cost of delay, and take chances.

—Brian Christian, Tom Griffiths

We are thankful to our *parents* and *families* for their boundless support through our life. No words can give them the right they deserve!!!

Special thanks to the Springer-publisher team, who showed us the ropes and gave us their thrust. We are highlight appreciating Prof. Amy Neustein, the series editor, for her support.

Last but not the least, we would like to thank our readers, hoping they will find the book as a valuable outstanding resource in their domain.

Nilanjan Dey Ph.D.
Amira S. Ashour Ph.D.

Contents

About the Authors

Nilanjan Dey was born in Kolkata, India, in 1984. He received his B.Tech. in Information Technology from West Bengal University of Technology in 2005, M.Tech. in Information Technology in 2011 from the same University, and Ph.D. in Digital Image Processing in 2015 from Jadavpur University, India.

In 2011, he was appointed as an Assistant Professor in the Department of Information Technology at JIS College of Engineering, Kalyani, India followed by Bengal College of Engineering College, Durgapur, India, in 2014. He is now employed as an Assistant Professor in the Department of Information Technology, Techno India College of Technology, India. His research topic is signal processing, machine learning, and information security.

He is an Associate Editor of IEEE Access and is currently the Editor-in-Chief of the International Journal of Ambient Computing and Intelligence, International Journal of Rough Sets and Data Analysis, Co-Editor-in-Chief of International Journal of Synthetic Emotion, International Journal of Natural Computing Research, Series Editor of Advances in Geospatial Technologies Book Series, and Co-Editor of Advances in Ubiquitous Sensing Applications for Healthcare (AUSAH) Elsevier (Book Series). Series Editor of Computational Intelligence in Engineering Problem Solving (CIEPS), CRC Press.

Amira S. Ashour was born in Tanta, Egypt, in 1975. She is graduated from Faculty of Engineering, Tanta University, Egypt, in 1997. She received her Master in Electrical Engineering in 2001 from the same university and Ph.D. in smart antenna in 2005 from the Department of Electronics and Electrical Communications Engineering, Faculty of Engineering, Tanta University, Egypt.

In 2005, she was appointed as a Lecturer in the Department of Electronics and Electrical Communications Engineering, Faculty of Engineering, Tanta University, Egypt. She was the Vice Chair of CS Department, CIT College, Taif University, KSA from 2009 till 2015. She was the Vice Chair of Computer Engineering Department, Computers and Information Technology College, Taif University, KSA for 1 year in 2015. She is now employed as an Assistant Professor and Head of Department in the Department of Electronics and Electrical Communications Engineering, Faculty of Engineering, Tanta University, Egypt. Her research topics are smart antenna, direction of arrival estimation, targets tracking, image processing, medical imaging, machine learning, and image analysis.

Abstract

Sensor array processing has various applications in speech processing, sonar, radar, seismology, and wireless communications. Speech sources' localization and Direction of Arrival estimation (DOAE) of radiating sensor arrays is considered a central signal processing research topic. DOA estimation systems receive the data from the sensor array in order to estimate the incoming signal's Direction of Arrival (DOA) for further localization of the speech source. Localization of the signal's source has been used in military location finding systems, in radar systems, in navigation, in tracking of several objects, and in various other applications including mobile communication systems. Sensor array processing has various applications in speech processing, sonar, radar, seismology, and wireless communications. Speech sources' localization and Direction of Arrival estimation (DOAE) of radiating sensor arrays is considered a central signal processing research topic. DOA estimation systems receive the data from the sensor array in order to estimate the incoming signal's Direction of Arrival (DOA) for further localization of the speech source. Localization of the signal's source has been used in military location finding systems, in radar systems, in navigation, in tracking of several objects, and in various other applications including mobile communication systems. Technological advancement in the fixed electronic devices, including teleconferencing and video systems as well as in the mobile electronic devices, including laptops and cell phones, increases the speech communication popularity in several contexts. Moreover, the increased communication demands between users require new services of better quality. Generally, blind handling of the microphone audio signals without prior knowledge of the signals has been developed to enhance the recorded speech. However, in order to improve the speech communication quality, it is essential to consistently determine the location of the speakers (speech source). Consequently, localization methods of speech/sound sources become the milestone for the speech enhancement methods that provide the sources' spatial information. Furthermore, the acoustic direction estimation problem in sonar is considered an open research area. High-resolution DOA estimation/localization algorithms and techniques become the main research area in array signal processing to track for example the mobile speech sources. In numerous audio/speech signal processing applications, DOAE of multiple mobile sound sources is a significant phase. This book is interested to support researchers, designers, and engineers in various interdisciplinary domains, such as engineering, speech processing, communication, direction of arrival estimation, and localization fields to ensure that the broad vision of the DOAE/localization of speech sources is well established. The book introduced the concept and model of the acoustics sources and models. Afterward, it highlights the most contemporary studies on this pervasive problem. The book provides a brief overview of the most classical direction of arrival estimation and localization techniques. In addition, employing the optimization algorithms to improve the DOAE techniques is also explored. The book highlighted the concept and principles of the multi-DOAE approaches. Using a microphone array, this book introduced the localization and tracking problem of multiple speech/acoustic sources.

Chapter 1
Introduction

Acoustics Science is concerned with the control, effects, production, transmission, and reception of sound. It is the study of mechanical waves in solid, liquid or gas in different media [1]. Acoustics include ultrasound, vibration, infrasound, speech, and sound. Acoustical engineering is concerned mainly with speech, such as speech processing, production, and transmission, detection, and perception [2–4]. For speech synthesis and recognition, the audio signal processing is carried out. Acoustics cover a massive range of areas, such as SONAR of submarine navigation, seismology, noise control, thermo-acoustic refrigeration, Bioacoustics, and ultrasound in medical imaging. Simultaneously, the speech communication is one of the significant categories under acoustics that models the speaker recognition, speech recognition, speech synthesis, oral dialogue, and language identification [5, 6]. It includes also a forensic speech science for speaker identification, forensic analysis of disputed utterances, and forensic voice comparison. With the technology evolution, peoples communicate more through cell phones, and tele-conferencing systems demanding high speech quality and new services and better speech quality.

Speech enhancement techniques of the microphones recorded speech and operate blindly deprived of any prior information of the signals [7, 8]. Such techniques are required with the spatial filtering of multiple microphones' signal that works blindly in the spatial domain. In order to improve the speech communication quality, the speakers' location has to be identified and localized accurately. In acoustic signal processing, several techniques are applied for audio beamforming and sound source localization [9–11]. These techniques are founded on capturing signals using microphone array for further array processing. Beamforming algorithms boost the audio signal received from the desired directions, though suppressed undesired signals from other directions [12–15]. Consequently, knowing the direction of sound source is essential for sound source localization. This process exploits the phase differences or time difference of arrival (TDOA) between the coming signals at each microphones' pair [16–18].

Dedicating source localization technique along with the speech enhancement approaches requires spatial information about the sources. In several applications,

© The Author(s) 2018
N. Dey and A. S. Ashour, *Direction of Arrival Estimation and Localization of Multi-Speech Sources*, SpringerBriefs in Speech Technology, https://doi.org/10.1007/978-3-319-73059-2_1

microphone arrays are implemented for speech recognition, dominant speaker's position located in the auditorium, and teleconferencing. Thus, DOAE of acoustic signals using a spatially separated set of microphones is essential in different practical applications, such as automatic steering cameras to the speaker in a conference room based on DOAE from microphones set [19–21].

Microphone arrays (sensor array) are designed for DOAE using the existing phase information in speech signals picked up by the microphones that are spatially alienated. Due to the spatial separation of the microphones, the acoustic speech signals reached them with different time of arrival. Time-delays (TD) of the received signals are estimated for each microphone pair in the array [22, 23]. Priory known array geometry is used for DOAE of the acquired signal using the TD measurements. The accurate DOA estimation is achieved from geometry and time-delays. The accuracy of the estimated DOA depends on several factors, namely sampling frequency, number of microphones, the used hardware for data acquisition, and the present noise in the captured signals.

The estimation of source location increases with the increased number of microphones in the array. For DOA estimate, typically about 10 till 40 microphones are used in the conventional microphone array systems [24–26]. Such large numbers of microphones necessitate the increase of channels for data acquisition, which sequentially increases the overall system cost. Exclusively, huge microphone arrays are required in several applications, such as automatic camera steering. Consequently, size and cost reduction is essential in acoustic source time-delay estimation as well as synchronized multiple data acquisition channels are required. In order to study the different DOAE techniques and localization of multi-speech sources scenarios, the types of the acoustics signals and systems are introduced.

Recently, signal processing is an energetic research domain in several applications, especially in communication, video conferencing, SONAR, RADAR, mobile communication, global positioning systems (GPS), sound source tracking, array processing, and speaker tracking [27, 28]. Localizing the source of speech/audio signal has received great attention of the researchers as an application of the signal processing methods. In such application, the signal processing methods are employed to discover the sound source direction that emits the signal. The speech processing system is concerned with finding the acoustic/sound source location (SSL) [29, 30]. The SSL system determines the sound (speech) source location relative to a reference frame.

In real life, humans discover the sound direction by using two ears, where the ears have different location compared to each other. The brain interprets the time delay between the two received signals to determine the incoming sound direction. This mechanism inspired the researchers to develop electronic systems to determine the speech/sound direction by using microphone sensors and to discover the source position.

Consequently, it is essential to study the fundamentals of the microphone arrays as capturing devices of the sound/speech signals before any further representation of the speech signals/sources DOAE techniques.

The organization of the remaining chapters is as follows. Chapter 2 includes an extensive description of the microphone array principles, including the acoustic signal and source models, the sensor array, the requirements of the speech processing systems, the far-field/near-field sources localizations, and the DOAE and localization of the speech sources. Chapter 3 introduced the different source localization and DOAE techniques for single as well as multiple sources. In addition, it includes the new trend of supporting the conventional DOAE techniques using the different optimization algorithms. Moreover, a brief highlights is directed toward the time of arrival estimation techniques. Chapter 4 provides three applied examples on the localization of multiple speech sources. Chapter 5 discussed in brief some challenges and new research openings in the speech direction of arrival estimation and localization. Finally, the book is concluded in Chap. 6.

References

1. Faahy, F. J. (2000). *Foundations of engineering acoustics*. London: Academic Press.
2. Gold, B., Morgan, N., & Ellis, D. (2011). *Speech and audio signal processing: Processing and perception of speech and music*. New York: Wiley.
3. Dunn, F., Hartmann, W. M., Campbell, D. M., & Fletcher, N. H. (2015). *Springer handbook of acoustics*. Berlin: Springer.
4. O'Shaughnessy, D. (2008). Automatic speech recognition: History, methods and challenges. *Pattern Recogn, 41*(10), 2965–2979.
5. Ververidis, D., & Kotropoulos, C. (2006). Emotional speech recognition: Resources, features, and methods. *Speech Commun, 48*(9), 1162–1181.
6. Junqua, J. C., & Haton, J. P. (2012). *Robustness in automatic speech recognition: Fundamentals and applications* (Vol. 341). Berlin: Springer Science & Business Media.
7. Butts, A. M. (2015). *Enhancing the perception of speech indexical properties of Cochlear Implants through sensory substitution*. Tempe, AZ: Arizona State University.
8. Hölig, C., Föcker, J., Best, A., Röder, B., & Büchel, C. (2014). Brain systems mediating voice identity processing in blind humans. *Hum Brain Mapp, 35*(9), 4607–4619.
9. Chen, J. C., Yao, K., & Hudson, R. E. (2002). Source localization and beamforming. *IEEE Signal Process Mag, 19*(2), 30–39.
10. Ward, D. B., Lehmann, E. A., & Williamson, R. C. (2003). Particle filtering algorithms for tracking an acoustic source in a reverberant environment. *IEEE Trans Speech Audio Process, 11*(6), 826–836.
11. Sheng, X., & Hu, Y. H. (2005). Maximum likelihood multiple-source localization using acoustic energy measurements with wireless sensor networks. *IEEE Trans Signal Process, 53*(1), 44–53.
12. Affes, S., & Grenier, Y. (1997). A signal subspace tracking algorithm for microphone array processing of speech. *IEEE Trans Speech Audio Process, 5*(5), 425–437.
13. Benesty, J., Chen, J., Huang, Y., & Dmochowski, J. (2007). On microphone-array beamforming from a MIMO acoustic signal processing perspective. *IEEE Trans Audio Speech Lang Process, 15*(3), 1053–1065.
14. Reuss, E. L., & Weeks, W. A. (2008). *U.S. Patent No. 7,359,504*. Washington, DC: U.S. Patent and Trademark Office.
15. Balanis, C. A., & Ioannides, P. I. (2007). Introduction to smart antennas. *Synthesis Lectures on Antennas, 2*(1), 1–175.

16. Valin, J. M., Michaud, F., & Rouat, J. (2007). Robust localization and tracking of simultaneous moving sound sources using beamforming and particle filtering. *Robotics and Autonomous Systems, 55*(3), 216–228.
17. Pertilä, P. (2009). *Acoustic source localization in a room environment and at moderate distances.*
18. Pavlidi, D., Griffin, A., Puigt, M., & Mouchtaris, A. (2013). Real-time multiple sound source localization and counting using a circular microphone array. *IEEE Trans Audio Speech Lang Process, 21*(10), 2193–2206.
19. Merimaa, J. (2002, April). Applications of a 3-D microphone array. In *Audio engineering society convention 112*. Audio Engineering Society.
20. Chang, P. S., Ning, A., Lambert, M. G., & Haas, W. J. (2002). *U.S. Patent No. 6,469,732.* Washington, DC: U.S. Patent and Trademark Office.
21. Huang, Y., Benesty, J., & Chen, J. (2006). Identification of acoustic MIMO systems: Challenges and opportunities. *Sig Process, 86*(6), 1278–1295.
22. Habets, E. A. P. (2007). Single- and multi-microphone speech dereverberation using spectral enhancement. *Dissertation Abstracts International, 68*(04).
23. Lavandier, M., & Culling, J. F. (2010). Prediction of binaural speech intelligibility against noise in rooms. *The Journal of the Acoustical Society of America, 127*(1), 387–399.
24. Hoshuyama, O., Sugiyama, A., & Hirano, A. (1999). A robust adaptive beamformer for microphone arrays with a blocking matrix using constrained adaptive filters. *IEEE Trans Signal Process, 47*(10), 2677–2684.
25. Dmochowski, J., Benesty, J., & Affes, S. (2007). Direction of arrival estimation using the parameterized spatial correlation matrix. *IEEE Trans Audio Speech Lang Process, 15*(4), 1327–1339.
26. Kumatani, K., McDonough, J., & Raj, B. (2012). Microphone array processing for distant speech recognition: From close-talking microphones to far-field sensors. *IEEE Signal Process Mag, 29*(6), 127–140.
27. Teixeira, T., Dublon, G., & Savvides, A. (2010). A survey of human-sensing: Methods for detecting presence, count, location, track, and identity. *ACM Comput Surv, 5*(1), 59–69.
28. Zekavat, R., & Buehrer, R. M. (2011). *Handbook of position location: Theory, practice and advances* (Vol. 27). New York: Wiley.
29. Zarghi, H. R., Sharifkhani, M., & Gholampour, I. (2011, December). Implementation of a cost efficient SSL based on an angular beamformer SRP-PHAT. In *2011 18th IEEE International Conference on Electronics, Circuits and Systems (ICECS)* (pp. 49–52). IEEE.
30. Cech, J., Mittal, R., Deleforge, A., Sanchez-Riera, J., Alameda-Pineda, X., & Horaud, R. (2013, October). Active-speaker detection and localization with microphones and cameras embedded into a robotic head. In *2013 13th IEEE-RAS International Conference on Humanoid Robots (Humanoids)* (pp. 203–210). IEEE.

Chapter 2
Microphone Array Principles

Microphone array (MA) involves several microphones positioned at diverse spatial locations. According to the sound propagation fundamentals, the multiple inputs can be handled to attenuate or to enhance the stemming signals from specific directions (desired signal) in the presence of demeaning noise sources [1, 2]. This process is based on the known source location information, and the used MA to guarantee hands-free signal as well as noise sturdiness. Traditional classic microphones should be near the speaker either by wearing the microphone or by using movable microphone. In order to solve such situation, microphone beamforming can be employed by using several microphones to generate directed beam patterns focusing on a specific speaker [3, 4]. These steerable microphones are motivated by the teleconferencing applications. In microphone array speech processing, a microphone array involves multiple microphones located at diverse spatial positions. Such arrays have an excessive impact in several speech processing, real-world applications, owing to their capability to deliver hands-free signal acquisition as well as noise robustness signals. According to the sound propagation fundamentals, the manifold inputs are handled to enrich the desired speech signals and to diminish the unwanted noise signals originating from other directions. This improvement of the input signal is founded on the information about the source location as well as the applied microphone array procedures.

MA is used to communicate acoustic signals, explicitly, speech signals. Array processing handles multiple sensors to transmit/receive the propagated signal waves [5]. These sensors have several applications as in microphone arrays to communicate the acoustic signals, such as speech signals. The sensor array is considered a new research area in speech processing that represents a sampled continuous apertures' version, where the aperture refers to a spatial area which transmits/receives the propagating waves [6]. The transmitting aperture is considered the active aperture, whereas the receiving aperture is a passive one. In acoustics such as speech, the aperture is an electro-acoustic transducer for converting the electrical signals into acoustic signals (loudspeaker) or vice versa (microphone). Inherently, the receiving aperture response is directional, where the detected signal amount by

© The Author(s) 2018

N. Dey and A. S. Ashour, *Direction of Arrival Estimation and Localization of Multi-Speech Sources*, SpringerBriefs in Speech Technology, https://doi.org/10.1007/978-3-319-73059-2_2

the aperture differs with the direction of arrival. Thus, the beam pattern/aperture directivity pattern is known as the aperture response, which is a function of the direction of arrival and frequency.

The sensor arrays' ultimate theory is applied in speech signal processing based on the wave propagation theory [7, 8]. In this context, the following array processing principles are considered. Sound waves broadcast as longitudinal waves through fluids, where in the propagation direction, the fluids' molecules move back/ forth leading to compression/expansion. For acoustic waves, the complex generalized wave equation depends on the fluid properties as a function of the sound pressure at a point in space and time. In addition, the propagation speed depends on the fluid density and pressure. For array processing algorithms, the spherical wave solution of the wave propagation equation indicates the dependency of the signal amplitude decay rate on the source distance, which has significant associations with sources in the near-field. Typically, sound waves are spherical in nature, however, at an adequate distance from the source, they can be considered as plane waves.

2.1 Models of the Acoustic Signals and Sources

In many applications, there are needs to design antennas with very high gains or directivity to meet the demands of long distance communication. By enlarging the dimensions of the antenna, high gains can be achieved. However, another way to achieve high gain without necessarily increasing the physical size of the individual elements will be to form an assembly of radiating elements arranged in a particular geometrical configuration [9]. This configuration of multi-elements is referred to as an antenna array. In most cases, the multi-elements in an antenna array are identical. Although not necessary, it is often simpler, more convenient and practical when it comes to the design and analysis of an antenna array. However, the primary reason for using antenna arrays is to be able to produce a directive beam that can be repositioned or scanned electronically.

2.1.1 Microphone Array

A phased array consists of multiple stationary radiating elements spaced a distance, d apart. The radiating elements have unity gain in all directions; i.e. they are isotropic radiators [10–12]. The radiating elements can be arranged in a few geometrical configurations, namely linear, circular and planar. Figure 2.1 illustrated the radiating elements different arrangement.

Consider M microphones in a linear microphone array with K incident signals from arrival angles q_k. Assume the incident waves to be plane waves, thus, each microphone in the array will collect each of the signals, s_k with time-delayed forms of each other. A reference point in the system can be considered to calculate each

(a)   **(b)**

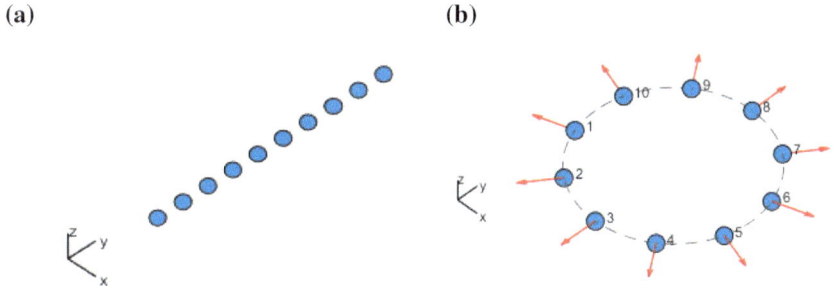

Fig. 2.1 Different microphones array configurations **a** linear array, and **b** circular array

Fig. 2.2 Microphone array
structure

signal phase. Figure 2.2 illustrated that microphone 1 is considered the reference
point, where d_i represented the distance from microphone m_1 to m_i. The planar
distance from source s_k to the microphone m_i that the sound travels relative to the
distance to reference microphone is given by [13, 14]:

$$distance_{i,k} = d_i cos\theta_k \qquad (2.1)$$

The equivalent time delay to each microphone in the array is represented by:

$$\tau_i = \frac{distance_{i,k}}{v} \qquad (2.2)$$

where v represents the sound velocity, which equals 343 m/s. At each microphone,
the signal s_k produces an input g_i as follows:

$$g_i(t) = |s_k(t - \tau_i)| \qquad (2.3)$$

The same $g_i(t)$ is produced for a source received from a direction between 0° and
−180°, where the source reached at its positive angle. For the narrowband cases,
this signal can be denoted as a phase shift of the received signal as follows:

$$g_i(t) = s_k(t)e^{-j\omega_0\tau_i} \qquad (2.4)$$

At the *ith* sensor, the complete received signal is a collection of the K incoming signals from K sources and the noise, which is given by:

$$g_i(t) = \sum_{k=1}^{K} s_k(t)e^{-j2\pi d_i \cos\theta_k/\lambda} + n_i(t) \tag{2.5}$$

where $\lambda = v/f_0$.

The microphones' spacing is an important issue that affects the microphone array system. Consequently, different microphone configurations can be tested to find the optimal configuration based on the problem under consideration [15]. Moreover, the spatial frequency domain considerations are crucial for avoiding aliasing, where smaller d/l leads to fewer side-lobes with nulls in the directional pattern. An omnidirectional pattern is formed as d/l tends to zero.

2.1.2   Near Field Considerations

For the incoming wavefront, most of the studies consider adaptive arrays having planar wave (far-field) hypothesis. However, in microphone arrays, the near-field hypothesis must be considered leading to spherical wave [16]. The near-field spherical wave has two main parameters, namely θ and r, where r is the radial distance from the source to the array. This complicated the model for solving the near-field problem.

Under the near-field hypothesis, the distance the traveling wave from the source to any point is equal to the line length that connects the two points [17]. Consequently, the distance r_i that the wave takes to each microphone equals the distance from the source to each microphone, where the relative time delay into each microphone can be established by the subtraction of the smallest r from each of r_1, r_2, r_3, r_4 with dividing the result by the sound speed.

2.1.3   Microphones Array Configurations

At different spatial locations, the microphones are placed in the microphone array that act as an omnidirectional acoustic antenna. Linear arrays, circular arrays and spherical microphone arrays are several configurations for designing the microphone array [18, 19]. For a source sound, the spatial location can be determined by the received signal correlation of the separated different microphones [20]. Mainly, the microphone array is used to enhance the signal under concern and to reduce/suppress the noise reduction using optimal filters for source separation as well as speakers tracking [21, 22]. Recently, the development of efficient speech communication devices has become an active research area with the advancement in the speech processing equipment [23, 24]. Such requirement is realized by microphone array that allows the user with hands free environment deprived of carrying microphone or speaking close to the microphone. Another application of the MA is in the video conferencing as well as for human machine interfaces and speech recognition.

Recently, MAs become popular due their accurate performance even with the limitations of the array processing (high hardware costs and large processing time). Researchers are interested to develop novel speaker localization algorithms to enhance the speech systems. These algorithms can be classified into one phase process, two phase process and spectral estimation [25, 26]. The different environmental conditions affect the performance of each algorithm. For example, room reverberation and various reflections of the received speech signal from a speaker in the acoustic environment lead to corrupt the signal from the background and additive noise. In order to resolve this problem, the speech signal can be recorded using a set of spatially MA, which needs localization as well as tracking of the moving speaker to steer electronically the MA to improve the quality of the speech acquisition. Moreover, speaker localization is essential in the multi-speaker scenarios [27, 28].

In order to perform such localization tasks different configurations of the MA can be used and selected according to the application under concern characteristics and parameters including the signal time variance, the spatial resolution, the acoustical situation, the frequency range, and the data processing [29, 30]. The MA is used to designate a complex sound field. Typically, the MA allows the signal capturing in several points at the same time with the cross spectra between each point to obtain the data presenting the direction of the coming sound at each time. MAs are used for signal recording that can be played back later in another space. Several microphones will help in understanding the sound field complex spatial pattern to renovate this pattern in another space.

2.1.4 Array Geometries

2.1.4.1 Scanning Arrays

As a rule of thumb, more microphones are required for better representation when back-propagating complex radiation pattern or when measuring a complex radiation field. However, increasing the number of microphones requires increasing the number of the recording channels. In order to solve this restriction, a scanning array can be used, where a small set of microphones are located at all desired positions to record the field. The total array configuration depends on the main microphones' mounting. For example, if the microphones are fixed in a linear array; they are appropriate in a regular grid spacing, while, if the microphones are fixed in a half-circle, the circle can be moved around a center point to attain a full sphere covered while the microphone recording process for static signals. A motor is used to automatically move the array around while the recording setup, consequently, it is conceivable to partially record the sound field at the different positions.

2.1.4.2 Planar Microphone Arrays

The planar array is considered the expected postponement to the linear array. Different configurations plane arrays (PAs) that consist of a regular microphones grid can be used for a simplified line array [31]. The Pas can be used in the complex radiating surface reconstruction.

2.1.4.3 Spherical Microphone Arrays

For a simple sound recording nearby a source, the spherical arrays are an expected choice, where the sound source is covered by a sphere in a regular angle, and the radiating body's radiation strength is directly measured. Spherical microphone array arrangements decompose the radiation into spherical harmonics. For the lowest harmonics, the same pressures occur at all microphones. Thus, in such case instead of using the spherical array, a dual-sphere array having two spheres of different radii can be employed. The optimal radius selection is considered a complex problem to be resolved before the recording.

The spherical arrays are considered as compact arrays with a small sphere in space instead of a bulky sphere nearby the radiator [32, 33]. In order to improve the traditional recording, the compact arrays that elaborated microphone recording configurations are considered. In such configuration, different orientation two directional microphones are used for recording.

Furthermore, in order to obtain a response equal that of the non-baffled directions, an omnidirectional microphone is applied. For example, using Matlab function "phased.OmnidirectionalMicrophoneElement" models the omnidirectional microphone that demands the operating frequency range specifications of the microphone [34]. In order to inspect the microphone directionality, namely the polar pattern, an azimuth cut is required by setting the elevation argument as zero single angle. Figure 2.3 both azimuth and elevation are from −90 to +90, which is illustrated the polar plot of a microphone power response at 20 Hz, 1 kHz, and 20 kHz; respectively, where Matlab 2017 is used to construct the omnidirectional microphone element with a response in the audible frequency range of the human, namely 20–20,000 Hz.

In Fig. 2.3, the green color represents the normalized power of '1', where all frequencies have the same response, which means that the microphone has a stable flat frequency response. Generally, in terms of the frequency response, there are two main microphone types, namely (i) flat frequency response, where all audible frequencies ranging from 20 Hz to 20 kHz and have the same output level, and (ii) tailored frequency response at which the microphone may have a peak in the frequency range from 2 to 8 kHz to increase intelligibility for live vocals. The flat

(a)

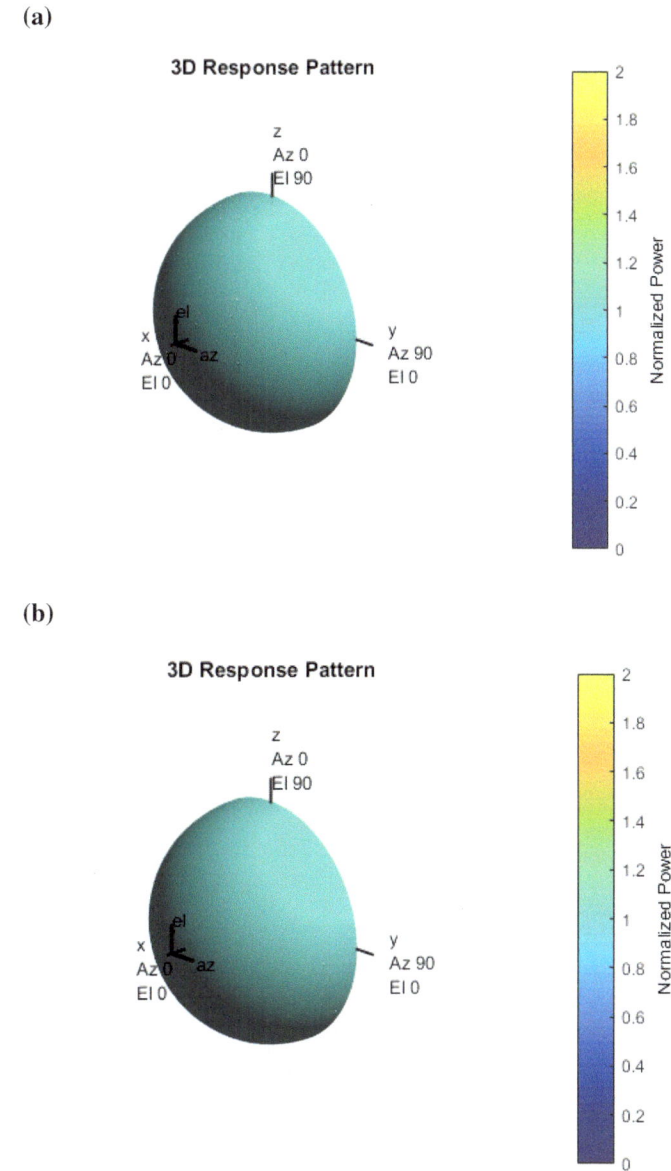

(b)

Fig. 2.3   a Three-direction (3D) power response pattern of omnidirectional microphone at 20 Hz. **b** 3D power response pattern of omnidirectional microphone at 1 kHz. **c** 3D power response pattern of omnidirectional microphone at 20 kHz

(c)

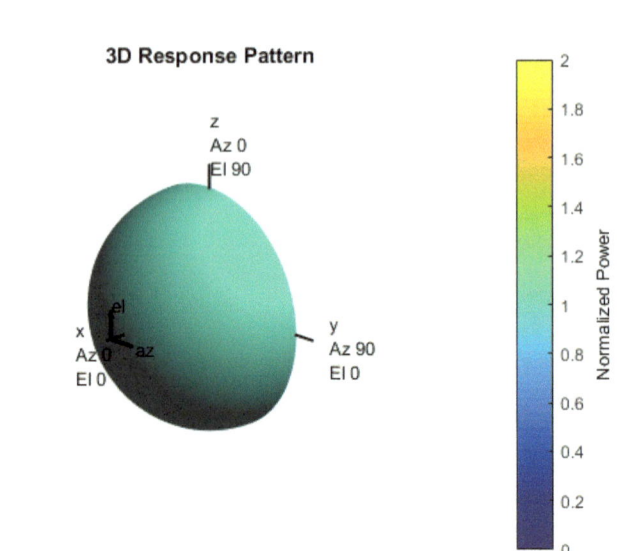

Fig. 2.3 (continued)

frequency response microphones are suitable for applications at which the sound source has to be reproduced (recording) without changing the original sound, while the tailored response microphones are ordinarily designed for the sound source enhancement in a precise application.

2.2 Sensor Arrays

The acoustic (speech) signals broadcast can be measured as functions of time/space variables with conserved signal information through the propagation. Accordingly, over all space and time, in order to renovate the signal it is required to temporally sample the band-limited signal at the specific space location or spatially sample the band-limited signal at the specific time instant. For all sensor/aperture array signal processing processes, the following associations are significant, namely (i) the dependence of the propagation speed on the medium characteristics, (ii) reinstating the signal over all space and time, (iii) the propagated wave decayed amplitude from the source, and (iv) propagating multiple waves prospect without interaction. For DOA estimation and source localization, separating the MAs signals using several algorithms can be engaged to discriminate the different signals inconsistent with their temporal and spatial characteristics, assuming lossless/homogeneous medium without considering the dispersion, propagation speed and diffraction.

The sampled continuous aperture is a discrete sensor array, where the aperture is excited at a limited number of discrete points. The overall sensor array's response is determined by the individual sensor response superposition that is consistent with the continuous aperture [35]. The equally spaced linear MA directivity pattern of identical sensors is contingent on the array's inter-element spacing, the frequency, and the number of array elements. The effective length of the sensor array is the continuous aperture length, while, the array actual physical length is signified by the distance between the first and last sensors in the array. In order to avoid aliasing and grating lobes in the directivity pattern, the sensor arrays implement spatial sampling with considering the Nyquist frequency. Moreover, the array gain is calculated to measure the sensor array effectiveness that represents the enhancement in signal-to-noise ratio between a reference sensor and the array output.

For a linear aperture, the wave source derives from the aperture far-field as plane waves with neglecting the wavefront curvature [36]. In several applied applications of the sensor arrays, such as speech recognition and DOAE, this criterion is unsatisfied as which the signal source is placed within the array near-field. In the near-field, in order to control a sensor array, the near-field directivity pattern is considered to match the far-field consistent directivity pattern of satisfying the frequency dependent sensor. Commendably, beam steering can be realized by applying time delays to the sensor inputs.

2.3 Speech Processing Requirements

Microphone arrays are the most advanced ways to improve the speech quality. A single microphone can provide the directivity for noise and reverberation reduction without any required post-processing [37]. However, an MA has been efficiently enhanced the speech signal quality by directing the received radiation pattern in the desired signal direction, thereby decreasing the interference and improving the captured sound quality. Figure 2.4 illustrated the MA that can be used for the speech signal enhancement in the direction of interest.

Fig. 2.4 Microphone arrays

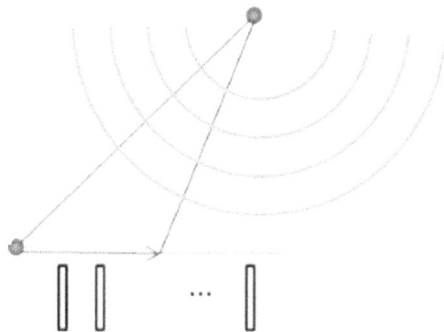

Array processing is the main speech processing requirements, where signals can be classified from the statistical point of view into deterministic/random wave-shape [38]. The deterministic wave refers to a known waveform that has unknown parameters, including the delay, amplitude, and/or scale. These waves are encountered in active sonar, data communications, and radar, where the transmitted signal waveform is known to the receiver. Nevertheless, in some applications, unknown signals are transmitted or signals may be affected by the transmission environment leading to changes with space and time. In this case the wave is considered random, where; all the obtained information is enclosed in the probability distribution function. However, in practice, the speech signals are disturbed by interference/noise that mix with the actual signal and modify its properties. Generally, interference has the same signal characteristics and may be produced by a similar source or due to the energy diffusion in a multiplex communication of one channel into the neighboring channel [39], electromagnetic coupling between two end-to-end wires, concurrent pick up of several stations by a single radio receiver, and the multipath transmission. The interference can be confused with the useful signal leading to its destructive nature, while the noise is commonly produced independent of the signal.

In array processing, the wave fields in space are signal-independent. A spherical isotropic noise is designed if the noise-field is produced by uncorrelated random waves propagating in all directions. Generally, the noise-field is assumed to be stationary in space and time. The information may convey by multi-dimensional signals, which are formed using frequency, time, or spatial diversity as in communication over a fading channel. The spatial field is sampled by multiple sensors in array processing. At different channels, the signals are collected from a multi-dimensional signal. The information combination of different dimensions can increase the processing performance. In communication, using a beamformer in array processing diversity diminishes the error reconstruction, and raises the signal-to-noise ratio (SNR).

The array processor goals are classified into field characterization or signal enhancement. Signal enhancement of the array output includes SNR improvement through beam steering in the source direction or through nulls inserting of the beam-pattern in the noise/interference direction [40]. The beam steering conventional methods places delay elements in the sensor output for further computing the delayed outputs' weighted sum. At the delay element's output, the signals having the same specific direction will appear at the same phase in the proper delays selection, which is known as beamforming. The array output signal power is amplified by the square magnitude of the number of sensors If the speech source is positioned at the beam direction. In the beamformer output, the noise power increases linearly with the increase in the sensors' number in an uncorrelated inter-sensor noise case. Consequently, the SNR increased by applying a conventional beamformer.

In order to estimate the source's spatial properties including the range, elevation, azimuth, velocity, and movement direction; the field characterization is required. The field characterization is completed in two stages, namely (i) determining the emitting source number, which is called detection, and (ii) localizing and estimating the signal position in the space. For the far field sources, the incoming wave fields are planar, thus, only azimuth and elevation directions can be estimated. Likewise, if the array and sources are in the same plane, the DOA will be the only spatial parameter of the emitting source.

In wireless communication, the narrow-band array processing is required, where a moving transmitter in indoor or mobile communication produces narrowband signals [41]. The receiver involves an antenna array that receives the original source's signal and its reflections from the nearby objects. In order to reduce the reflected wavefronts effect, the antenna array steered a beam in the source direction to estimate the source location. The antenna array drives the power in the transmission mode in the source direction only by establishing a steered beam. Thus, the energy is preserved and meanwhile the power is transmitted only in a specific direction where smaller interfering influences the other receivers.

Array processing procedures can similarly be realistic with wideband signals, where the frequency bandwidth is comparatively large associated with the center frequency [42]. A microphone array for instance, can be used to localize a speaker in a specific place. At the array, the arriving signal is a wideband speech signal accompanied by reflections from the surroundings. The reflection effect that interferes with the direct signal can be compensated by beam steering towards the speaker direction. Another wideband array processing example is the MA can be used in the hands-free mobile communications inside a car, where the driver voice is collected by microphone array. Forming a beam-pattern with nulls in the noise direction reduces the car environment noise. The foremost noise fields in the microphone array applications are considered according to the correlation degree between the noise signals at dissimilar spatial locations. The coherence is considered the most used measurements of the correlation. The directly propagated signal noise to the microphones is deliberated as coherent noise without any form of dispersion, dissipation, or reflection due to the acoustic environment. Conversely, the measured noise at any certain spatial location in an incoherent noise field is uncorrelated with the measured noise at all other locations. Generally, in the MA applications, the electrical noise in the microphones is distributed randomly, which can be considered an incoherent noise source. A diffuse noise field refers to noise of identical energy that broadcasts simultaneously in all directions. Therefore, in the diffuse noise field, the sensors will receive noise signals, which are poorly correlated, but have the same energy [43]. Numerous practical noise situations can be branded by a diffuse noise field, including the car noise. In a diffuse noise field, the noises coherence at any two points is a function of the distance between the sensors.

2.4 Microphone Array Beamforming

Beamforming techniques can be categorized according to the data type, namely data-dependent/data-independent. In the data-dependent (adaptive beamforming) methods are continuously adapted their parameters based on the received signals

[44]. Nevertheless, the data-independent (fixed beamforming) methods have beamformers with fixed parameters during the operation. For different noise situations, different beamforming methods can be proper to determine the encountered noise fields in the MA applications.

For speech applications, MAs have a small number of elements that should cover only the audible range of the electromagnetic spectrum. Consequently, compared to other applications, such as wireless beamforming, the precise array manifold can be pre-designed. The beamforming MA techniques are generally belongs to signal processing, which an extensive range of applications has extended from source detection/separation to de-noising or de-reverberation that have the common delay and filter methods. These approaches intend to improve the speech signal using a multichannel recording. Hence, these signal reconstruction converges to effective results using four to eight microphones. Various filtering methods are based on the speech processing, including Wiener filter, and Kalman filter [45]. Typically, the extra sources are correlated between the microphones without any correlation with the source under concern. In addition, the reverberation is associated between the microphones as well as with the source of interest. Such unwanted signals should be separately treated.

Direction-of-Arrival (DOA) estimation has an energetic role in several applications, especially for speakers' localization. Beamforming is the most protuberant procedure for estimating the DOA. It is used to accompany by sensors/antenna array to transmit/receive signals to/from certain spatial direction even with the incidence of noise/interference. One of the simplest microphone array beamforming methods is the delay-sum beamforming (DS). In order to steer the directivity pattern's main lobe in the wanted direction, the phase weights apply to the input channels. In the delay-sum beamforming, the sensor inputs in the time domain are mainly delayed by specific seconds, and then a summation process is applied to provide a single array output. In the summation, each channel is prearranged an equal amplitude weighting for unity directivity pattern in the desired direction. The DS is considered one of the filter sum beamformers at which both the amplitude/phase weights are frequency dependent. Several beamformers are considered filter-sum beamformer that use matrix algebra to describe the microphone array methods.

The speech signal is broadband, thus, for frequency invariant beam-pattern; a single linear array design is insufficient [46]. In order to handle the speech broadband signals, an array can be implemented as a series of linear sub-arrays with uniform spacing to provide the desired response characteristics for a specific frequency range. Smaller array length is essential with high frequencies to preserve constant beam-width. Furthermore, in each sub-array, the number of elements must remain the same to guarantee the same side-lobe level across different frequency bands. Thus, the sub-arrays are realized in a nested manner, where any sensor can be used in several sub-arrays.

Super-directive beamforming is another beamforming procedure that achieves the ability for closely spaced end-fire arrays. Its channel filters are expressed to maximize the array gain [47]. Its filters depend only on the array geometry and the source location that are calculated for a specified configuration. The super-directive techniques are efficient as it provides satisfactory array performance at low frequencies for speech processing applications.

For conventional beamforming procedures, low frequency performance is challenging due to the large wavelengths that provide insignificant phase differences between closely spaced sensors that lead to deprived directive discrimination. Near-field super-directivity is an adjustment of the typical super-directive method that considers the amplitude differences and the phase differences [48]. It provides directional sensitivity with greater performance compared to the standard near-field for sources at low frequencies, where the phase differences are insignificant at low frequencies, mainly when the sensors are positioned in an end-fire arrangement, which maximizes the distance difference from the source to each microphone.

2.5 Far-Field and Near-Field Source Location

In several engineering applications, source localization is the vital stage in speech signal processing, array signal processing, and wireless communication. Different DOAE estimation algorithms are developed for far field narrow band sources. However, it becomes intricate in the case of near-field sources in the Fresnel zone of the array aperture. In near-field scenarios, such as the speech enhancement with microphone array in a conference room, seismic exploration; under water source localization; and ultrasonic imaging, the impinging wavefront on the array is spherical, thus, range information is required along with the estimated DOA of the sources for accurate localization. Researchers are interested to develop methods for near-field source localization, such as the two-dimensional (2D) Multiple Signal Classification (MUSIC) method, Maximum Likelihood (ML) method, high-order spectra (HOS) based algorithms, and Estimation of Signal Parameters via Rotational Invariance Technique (ESPRIT) method [49, 50]. In practical situations, especially in the case of closely spaced, multiple sources, these techniques are computationally heavy and needs extra computations to pair the parameters, which causes poor DOAE in low Signal to Noise Ratio (SNR).

2.6 Speech Source Direction of Arrival Estimation and Localization

Microphone arrays are essential for spatial analysis, where numerous microphones perceive more than only one microphone. The speech analysis and reconstruction become easier with more collected information on the space [51]. Microphone array methods are involved in several applications that can be categorized into two core domains according to their problems and mathematical models, namely (i) the beamforming methods are based on signal processing, and (ii) the recorded sound field back-propagation against the radiating body according to the Helmholtz–Kirchhoff integral.

Complex sound/speech environments, such as classrooms/auditoriums, cars, planes, trains, and outdoor places, depend on the listener/speaker position, where any change in the location leads to a corresponding change in the sound perception. Thus, it is essential to determine distance to the source estimation, the source location at a definite angle, separating the different speakers, concentrating with certain speaker; which called cocktail party effect. For example, the position of musical instruments in musical space facilities the separation process of them. The height, depth, or size of a space can be estimated using reverberation information. For speech extraction and speaker selection, some reverberation can be neglected.

The standard beamforming can be considered as a source localization problem that can be enhanced by using several microphones recording from different positions [52]. Such a case occurs in the reverberant situations at which the source is reflected several times, and the source position localization become complex compared to free space. Moreover, this problem becomes more complex in the case of multiple sources, such as many speakers in a room. In conference rooms and in a car, the noise and reverberation are strong, thus, it is essential to localize the source as well as to use the MA to de-noise the recorded signal. In addition, de-reverberate the signal and reconstruct the speech signals are vital. These requirements are achieved through using fixed array to measure the reverberation easily to increase speech intelligibility.

2.6.1   Sound/Speech Source Localization

Localizing single sound source at a distance is considered a near-field problem. In contradiction of the equivalent source procedures that produce a complex radiation pattern, the localization estimation produces a single angle value relative to the normal direction of the MA, which is the preferred single source location [53]. The phase of a signal $s(t)$ is considered the most significant property for its detection. At each of N microphones, the signal phase is different, recording a time series $m_i(t)$, where $i = 1, 2, 3, \ldots N$:

$$m_i(t) = A_i s(t + \Delta t_i) + \xi_i(t) \tag{2.6}$$

where $\xi_i(t)$ is the uncorrelated noise between the microphones and Δt_i are the different time delays between the source point and the i microphones. When a signal traveling from the far field reaches at the array, the arrival time difference of this wave at the microphones [54] can be expressed as:

$$\Delta t_{ij} = \Delta t_i - \Delta t_j = \frac{r_i - r_j}{c} \tag{2.7}$$

where r_i and r_j are the distances between the source and its corresponding micro-phones as c is the sound speed.

Assume a source located at specific DOA with the angle φ_i between the source and the MA plane at microphone i. Using three microphones of distance d, thus,

$$r_2^2 = r_1^2 + d^2 + 2r_1 d cos(\varphi_1) \qquad (2.8)$$

$$r_3^2 = r_1^2 + 4d^2 + 4r_1 d cos(\varphi_1) \qquad (2.9)$$

For three microphones, the time delay between a sensor i and its nth neighbour is:

$$\Delta t_i = \frac{(n-1)d}{c} cos(\varphi) \qquad (2.10)$$

Consequently, the time delay Δt_i can be estimated from the phase difference between the two sensors and then the DOA φ can be measured. Moreover, once the angle is identified, the distances r_i between the source and sensors can be estimated.

2.6.2   Directional of Arrival Estimation

The angle of arrival or the wave number estimation problem of a plane wave is considered as DOA estimation problem or direction finding. DOAE plays an energetic role in several applications including sonar, radar, electronic surveillance, seismic systems, radio astrology, and medical diagnosis. Beamforming is consid-ered the most protuberant method for estimating the DOA. Over the last several decades, the DOAE has magnetized the researchers' attention due to its widespread applications and the complexity of determining the optimum estimator. Numerous approaches were developed to address the DOAE problem of multiple sources using the received signals at the sensors. Array processing requires either the knowledge of the direction of the desired signal source or a reference signal. Thus, antenna arrays are extensively used to resolve the direction finding problems.

For the sound source, the DOA is imperative information for any beamforming system. The beamformer can direct itself to capture signals coming only from the DOA, while ignoring others [55]. Time delay of arrival is considered one of the most prevalent DOAE approaches, which is called phase differencing. For esti-mating the DOA delay, consider a sound source located in the far field, two microphones on the same plane will receive the signal with slightly different times. The precise arrival difference value is based on the sound source angle relative to the microphones. Hence, the delay information can be used for estimating the direction at which the source is impinging on the array for further measuring of the propagation delay between the microphones to steer the beamformer toward the captured signals in the desired direction and rejecting all other signals [56].

References

1. Levy, A., Gannot, S., & Habets, E. A. (2011). Multiple-hypothesis extended particle filter for acoustic source localization in reverberant environments. *IEEE Transactions on Audio, Speech and Language Processing, 19*(6), 1540–1555.
2. Doclo, S., Kellermann, W., Makino, S., & Nordholm, S. E. (2015). Multichannel signal enhancement algorithms for assisted listening devices: Exploiting spatial diversity using multiple microphones. *IEEE Signal Processing Magazine, 32*(2), 18–30.
3. Ryan, J. G., & Goubran, R. A. (2000). Array optimization applied in the near field of a microphone array. *IEEE Transactions on Speech and Audio Processing, 8*(2), 173–176.
4. Benesty, J., Chen, J., & Huang, Y. (2008). *Microphone array signal processing* (Vol. 1). Berlin: Springer Science & Business Media.
5. Haimovich, A. M., Blum, R. S., & Cimini, L. J. (2008). MIMO radar with widely separated antennas. *IEEE Signal Processing Magazine, 25*(1), 116–129.
6. Huang, K., & Larsson, E. (2013). Simultaneous information and power transfer for broadband wireless systems. *IEEE Transactions on Signal Processing, 61*(23), 5972–5986.
7. Chan, S. C., & Chen, H. H. (2007). Uniform concentric circular arrays with frequency-invariant characteristics—theory, design, adaptive beamforming and DOA estimation. *IEEE Transactions on Signal Processing, 55*(1), 165–177.
8. Gorodetskaya, E. Y., Malekhanov, A. I., Sazontov, A. G., & Vdovicheva, N. K. (1999). Deep-water acoustic coherence at long ranges: Theoretical prediction and effects on large-array signal processing. *IEEE Journal of Oceanic Engineering, 24*(2), 156–171.
9. Legay, H., & Shafai, L. (1994). New stacked microstrip antenna with large bandwidth and high gain. *IEE Proceedings-Microwaves, Antennas and Propagation, 141*(3), 199–204.
10. Rodenbeck, C. T., Li, M. Y., & Chang, K. (2003). A novel millimeter-wave beam-steering technique using a dielectric-image line-fed grating film. *IEEE Transactions on Antennas and Propagation, 51*(9), 2203–2209.
11. Krnac, B., & Haring, J. (2008, April). Improving of UHF RFID pattern steering by phased antenna array. In *2008 18th International Conference Radioelektronika* (pp. 1–4). IEEE.
12. Cheng, D. K. (1971). Optimization techniques for antenna arrays. *Proceedings of the IEEE, 59*(12), 1664–1674.
13. Brandstein, M., & Ward, D. (Eds.). (2013). *Microphone arrays: Signal processing techniques and applications*. Berlin: Springer Science & Business Media.
14. Brandstein, M. S., Adcock, J. E., & Silverman, H. F. (1995). A practical time-delay estimator for localizing speech sources with a microphone array. *Computer Speech & Language, 9*(2), 153–169.
15. Boone, M. M., Cho, W. H., & Ih, J. G. (2009). Design of a highly directional endfire loudspeaker array. *Journal of the Audio Engineering Society, 57*(5), 309–325.
16. Dmochowski, J. P., Benesty, J., & Affes, S. (2007). A generalized steered response power method for computationally viable source localization. *IEEE Transactions on Audio, Speech and Language Processing, 15*(8), 2510–2526.
17. Rumsey, F., & McCormick, T. (2012). *Sound and recording: An introduction*. Boca Raton, FL: CRC Press.
18. Alexandridis, A., Griffin, A., & Mouchtaris, A. (2017). *U.S. Patent No. 9,549,253*. Washington, DC: U.S. Patent and Trademark Office.
19. Roig, E. T. (2014). *Eigenbeamforming array systems for sound source localization* (Doctoral dissertation, Ph.D. thesis). Technical University of Denmark.
20. Zhu, N. (2011). *Locating and extracting acoustic and neural signals* (Doctoral dissertation). Wayne State University.
21. Brandstein, M., & Ward, D. (Eds.). (2013). *Microphone arrays: Signal processing techniques and applications*. Berlin: Springer Science & Business Media.

22. Valin, J. M., Rouat, J., & Michaud, F. (2004, September). Enhanced robot audition based on microphone array source separation with post-filter. In *2004 IEEE/RSJ International Conference on Intelligent Robots and Systems, 2004 (IROS 2004). Proceedings* (Vol. 3, pp. 2123–2128). IEEE.

23. Perez-Meana, H. (Ed.). (2007). *Advances in audio and speech signal processing: Technologies and applications.* Hershey: Igi Global.

24. Huang, X., Baker, J., & Reddy, R. (2014). A historical perspective of speech recognition. *Communications of the ACM, 57*(1), 94–103.

25. Markovich, S., Gannot, S., & Cohen, I. (2009). Multichannel eigenspace beamforming in a reverberant noisy environment with multiple interfering speech signals. *IEEE Transactions on Audio, Speech and Language Processing, 17*(6), 1071–1086.

26. Kumar, L., Tripathy, A., & Hegde, R. M. (2014). Robust multi-source localization over planar arrays using music-group delay spectrum. *IEEE Transactions on Signal Processing, 62*(17), 4627–4636.

27. Ma, W. K., Vo, B. N., Singh, S. S., & Baddeley, A. (2006). Tracking an unknown time-varying number of speakers using TDOA measurements: A random finite set approach. *IEEE Transactions on Signal Processing, 54*(9), 3291–3304.

28. Minotto, V. P., Jung, C. R., & Lee, B. (2014). Simultaneous-speaker voice activity detection and localization using mid-fusion of SVM and HMMs. *IEEE Transactions on Multimedia, 16* (4), 1032–1044.

29. Brutti, A., & Nesta, F. (2013). Tracking of multidimensional TDOA for multiple sources with distributed microphone pairs. *Computer Speech & Language, 27*(3), 660–682.

30. Plinge, A., Jacob, F., Haeb-Umbach, R., & Fink, G. A. (2016). Acoustic microphone geometry calibration: An overview and experimental evaluation of state-of-the-art algorithms. *IEEE Signal Processing Magazine, 33*(4), 14–29.

31. Rafaely, B. (2005). Analysis and design of spherical microphone arrays. *IEEE Transactions on Speech and Audio Processing, 13*(1), 135–143.

32. Li, Z., & Duraiswami, R. (2007). Flexible and optimal design of spherical microphone arrays for beamforming. *IEEE Transactions on Audio, Speech and Language Processing, 15*(2), 702–714.

33. Balmages, I., & Rafaely, B. (2007). Open-sphere designs for spherical microphone arrays. *IEEE Transactions on Audio, Speech and Language Processing, 15*(2), 727–732.

34. https://www.mathworks.com/help/phased/ug/omnidirectional-microphone.html#btdemjn.

35. Zaunschirm, M. (2012). *Modal beamforming using planar circular microphone arrays* (Doctoral dissertation, MS thesis). University of Music and Performing Arts Graz.

36. Duan, K., & Lü, B. (2003). Nonparaxial analysis of far-field properties of Gaussian beams diffracted at a circular aperture. *Optics Express, 11*(13), 1474–1480.

37. Habets, E. A. P. (2007). Single- and multi-microphone speech dereverberation using spectral enhancement. *Dissertation Abstracts International, 68*(04).

38. Hickok, G., & Poeppel, D. (2007). The cortical organization of speech processing. *Nature Reviews Neuroscience, 8*(5), 393–402.

39. Carrasco, J. M., Franquelo, L. G., Bialasiewicz, J. T., Galván, E., PortilloGuisado, R. C., Prats, M. M., ... & Moreno-Alfonso, N. (2006). Power-electronic systems for the grid integration of renewable energy sources: A survey. *IEEE Transactions on industrial electronics, 53*(4), 1002–1016.

40. Omologo, M., Matassoni, M., & Svaizer, P. (2001). Speech recognition with microphone arrays. In *Microphone arrays* (pp. 331–353). Berlin: Springer.

41. Jensen, M. A., & Wallace, J. W. (2004). A review of antennas and propagation for MIMO wireless communications. *IEEE Transactions on Antennas and Propagation, 52*(11), 2810–2824.

42. Kilfoyle, D. B., & Baggeroer, A. B. (2000). The state of the art in underwater acoustic telemetry. *IEEE Journal of Oceanic Engineering, 25*(1), 4–27.

43. Saligrama, V., Alanyali, M., & Savas, O. (2006). Distributed detection in sensor networks with packet losses and finite capacity links. *IEEE Transactions on Signal Processing, 54*(11), 4118–4132.
44. Li, J., & Stoica, P. (2005). *Robust adaptive beamforming* (Vol. 88). New York: Wiley.
45. Chen, J., Benesty, J., Huang, Y., & Doclo, S. (2006). New insights into the noise reduction Wiener filter. *IEEE Transactions on Audio, Speech and Language Processing, 14*(4), 1218–1234.
46. Brandstein, M., & Ward, D. (Eds.). (2013). *Microphone arrays: Signal processing techniques and applications*. Berlin: Springer Science & Business Media.
47. Benjamin, R. (1983). Signal-space, the key to signal processing. *Radio and Electronic Engineer, 53*(11), 407–422.
48. Tang, M. C., & Ziolkowski, R. W. (2013). Efficient, high directivity, large front-to-back-ratio, electrically small, near-field-resonant-parasitic antenna. *IEEE Access, 1,* 16–28.
49. Sheikh, Y. A., Ullah, R., & Zhonfu, Y. (2016). Range and direction of arrival estimation of near-field sources in sensor arrays using differential evolution algorithm. *International Journal of Computer Applications (IJCA), 139*(4), 16–20.
50. Waweru, N. P., Konditi, D. B. O., & Langat, P. K. (2014). Performance analysis of MUSIC, root-MUSIC and ESPRIT DOA estimation algorithm. *International Journal of Electrical, Computer, Energetic, Electronic and Communication Engineering, 8*(1), 209–216.
51. Fristrup, K. M., & Mennitt, D. (2012). Bioacoustical monitoring in terrestrial environments. *Acoustics Today, 8*(3), 16–24.
52. O'Donovan, A., Duraiswami, R., & Neumann, J. (2007, June). Microphone arrays as generalized cameras for integrated audio visual processing. In *IEEE Conference on Computer Vision and Pattern Recognition, 2007. CVPR'07* (pp. 1–8). IEEE.
53. Gotsis, K. A., Siakavara, K., & Sahalos, J. N. (2009). On the direction of arrival (DoA) estimation for a switched-beam antenna system using neural networks. *IEEE Transactions on Antennas and Propagation, 57*(5), 1399–1411.
54. DiBiase, J. H., Silverman, H. F., & Brandstein, M. S. (2001). Robust localization in reverberant rooms. In *Microphone arrays* (pp. 157–180). Berlin: Springer.
55. Zhao, F., & Guibas, L. J. (2004). *Wireless sensor networks: An information processing approach*. Los Altos, CA: Morgan Kaufmann.
56. Valin, J. M., Michaud, F., Rouat, J., & Létourneau, D. (2003, October). Robust sound source localization using a microphone array on a mobile robot. In *2003 IEEE/RSJ International Conference on Intelligent Robots and Systems, 2003 (IROS 2003). Proceedings* (Vol. 2, pp. 1228–1233). IEEE.

Chapter 3
Sources Localization and DOAE Techniques of Moving Multiple Sources

Estimating the direction of non-stationary moving array as well as moving narrowband sources is considered an active research area. The most cutting-edge techniques are originated from the maximum likelihood (ML) and expectations maximization (EM) methods as they have a form of recursive extended Kalman filters and use built-in source-movement models. For non-stationary speech sources, nonparametric modeling of the source movements can be employed. Such models have no mathematical model of the assumed signal, while, parametric approaches have a mathematical model to define the signal form and to estimate it [1].

Consider the problem of localizing q speech sources by using an array of n passive sensors. In order to obtain the signal model, where the sources generate a wave-field traveling through space and sampled by the sensor array. The array aperture is the space occupied by the array and usually is measured in units of signal wavelength. For point sources (omnidirectional) in the far field of the array, the only parameter that characterizes a difference of the signals impinging on the sensors from a source is a time-delay, which is called angle of arrival (AOA), or direction of arrival (DOA).

For a uniform linear array, consider the array of n identical sensors uniformly spaced on a line that receives $q(q < n)$ narrowband signals impinging from the unknown varying directions $\{\theta_1, \ldots, \theta_q\}$. The $n \times 1$ output vector of the array at the discrete time t is modeled as [2–7]:

$$\mathbf{r}(t) = \mathbf{A}(t)\mathbf{s}(t) + \mathbf{e}(t) \tag{3.1}$$

where, the $n \times q$ time-varying direction matrix is given by:

$$\mathbf{A}(t) = [\mathbf{a}(\theta_1(t)), \mathbf{a}(\theta_2(t)), \ldots, \mathbf{a}(\theta_q(t))] \tag{3.2}$$

The $\mathbf{A}(t)$ matrix is composed of the source direction vectors, which are known as the steering vectors that defined as follows.

© The Author(s) 2018
N. Dey and A. S. Ashour, *Direction of Arrival Estimation and Localization of Multi-Speech Sources*, SpringerBriefs in Speech Technology,
https://doi.org/10.1007/978-3-319-73059-2_3

$$\mathbf{a}(\theta_i) = \left[1, \exp(-j\frac{2\pi}{\lambda}d\sin\theta_i), \dots, \exp(-j(n-1)\frac{2\pi}{\lambda}d\sin\theta_i)\right]^T \qquad (3.3)$$

where $i = 1, \dots, q$, λ is the wavelength defined as the distance traveled by the harmonic carrier signal in one period; d is the equi-spaced inter-element spacing; $\mathbf{s}(t)$ is a $q \times 1$ vector of the source waveforms; $\mathbf{e}(t)$ is $n \times 1$ vector of sensor noise as the white zero-mean Gaussian random noise with a variance of σ^2 that has:

$$E\{\mathbf{e}(t)\} = 0, \; E\{\mathbf{e}(t)\mathbf{e}(t)^H\} = \sigma^2 I, E\{\mathbf{e}(t)\mathbf{e}(t)^T\} = 0 \qquad (3.4)$$

where $E\{\}$ means the expectation, $(.)^H$ refers to the Hermitian conjugation and $(.)^T$ stands for the transpose.

After setting this model, the source location problem is turned into a time-invariant or time-variant parameter estimation problem.

3.1   Direction of Arrival Estimation Techniques

In many cases, the receiver cannot determine which direction a speech signal will arrive from. Accordingly, the DOA estimation step becomes essential before beam-forming. The array-based DOA estimation techniques can be broadly divided into conventional, subspace-based, maximum likelihood, and integrated techniques [8].

3.1.1   Conventional Beamformer for DOAE

Conventional beamformer approaches are based on the concepts of beamforming and null steering without exploiting the nature of the received signal model or the statistical model of the signals/noise. These beamformers are electronically steer the beams in the possible directions and look for peaks in the output power. The delay-and-sum methods are considered classical beamformers for DOA estimation. However, these methods suffer from poor resolution, where the width of the beam and the height of the side lobes limit the effectiveness when the signals are from multiple sources. Capon's Minimum Variance method tries to overawe the poor resolution problems related to the delay-and-sum technique. Capon's method has several disadvantages, namely (i) it fails in the presence of other signals that are correlated with the Signal-of-Interest (SOI), and (ii) it is expensive for large arrays, where it requires the computation of matrix inverse.

The classical beamforming based methods have fundamental limitations in resolution. These limitations arise due to the neglecting of the input data model structure. Generally, these conventional methods need a large number of elements to accomplish high resolution. A conventional beamformer as the delay-and-sum beamformer selects the phases to steer the array in a particular direction, known as the look direction [3]. It has been introduced as a natural extension of the standard Fourier-based spectral analysis of the sensor array data. The model of a finite impulse response spatial filter with the output for the signal model is given by:

$$y(t) = \sum_{k=1}^{n} \omega_k^* r_k(t) = \mathbf{W}^H \mathbf{r}(t), \quad \mathbf{W} = (\omega_1, \ldots \ldots, \omega_n)^T \tag{3.5}$$

Assume that the waveform $\mathbf{s}(t)$ and the noise $\mathbf{e}(t)$ are zero-mean independent random processes. Given samples $\mathbf{r}(t)$, thus, the output power is measured by:

$$P(\mathbf{W}) = \frac{1}{N} \sum_{t} |y(t)|^2 = \mathbf{W}^H \hat{\mathbf{R}}_r \mathbf{W} \tag{3.6}$$

Design a spatial filter suppressing the noise component and preserving the waveform signal $\mathbf{s}(t)$. It can be done by the following optimization problem [4]:

$$\min_{\mathbf{W}} \|\mathbf{W}\|^2 \text{ subject to } \mathbf{W}^H \mathbf{a}(\theta) = 1 \tag{3.7}$$

The optimal weights vector for the spatial filter is expressed by:

$$\mathbf{W} = \mathbf{a}(\theta)/n \tag{3.8}$$

Substitute by these weights, the output power of the spatial filter as a function of θ is obtained as follows:

$$P_{conv}(\theta) = \frac{1}{n^2} \mathbf{a}^H(\theta) \hat{\mathbf{R}}_r \mathbf{a}(\theta) \tag{3.9}$$

In order to find θ of the unknown DOA, the power $p_{conv}(\theta)$ is maximized on θ using the following expression:

$$\hat{\theta} = \arg(\max_{\theta} P_{conv}(\theta)) \tag{3.10}$$

Several methods including Capon's beamforming, MUSIC and other subspace-based methods use the power function analysis as a basic original idea for the development. However, one can start from the observation model to obtain the following residual function:

$$I(\theta, s) = \frac{1}{N} \sum_{t} \|\mathbf{r}(t) - \mathbf{a}(\theta)s(t)\|^2 \tag{3.11}$$

The quadratic residual function gives the maximum likelihood estimates of the angle θ and the waveform $s(t)$ provided that the random errors are Gaussian [9]. This beamformer deals with stationary sources only, so other beamformers can be used to handle the non-stationary sources.

3.1.2 Subspace DOA Estimation Methods

It is well known that several super resolution approaches are available for estimating the DOA of signals received by array including MUSIC, and ESPRIT. All are eigenvalue decomposition problems. The difference between these algorithms is in how the information is used to determine the DOA.

The MUSIC algorithm is a high resolution technique that provides information about the number of incident signals, signal DOA, noise power, etc. It can resolve closely spaced signals that cannot be detected by Capon's method. In the MUSIC algorithm, an exhaustive search is performed looking for the signals that are orthogonal to the noise subspace. Various modifications of the MUSIC algorithm have been proposed to decrease the computational complexity and increase its resolution performance. These modified versions include (i) the Root-MUSIC algorithm, which is based on the polynomial rooting and provides higher resolution. It reduces the number of calculations by avoiding an exhaustive search [10], however, it is applicable only to a uniformly spaced linear array. In addition, (ii) the cyclic MUSIC which is a signal selective direction finding algorithm that exploits the spectral coherence of the received signal as well as the spatial coherence to improve the performance of the conventional MUSIC algorithms. By exploiting spectral correlation along with MUSIC, it is possible to resolve signals spaced more closely than the resolution threshold of the array when only one of them is the SOI [11].

In this connection, it may be mentioned that the ESPRIT algorithm is another subspace-based DOA estimation technique that reduces the computational and storage requirements of MUSIC and does not involve an exhaustive search through all possible steering vectors to estimate the DOA. Unlike MUSIC, ESPRIT does not require prior knowledge about the array manifold vectors.

3.1.3 Maximum Likelihood Techniques

Maximum likelihood (ML) techniques are considered effective techniques for DOA estimation, which are computationally intensive. The ML methods are better than the subspace based methods, expressly in SNR conditions or in the case of small number of signal samples [12]. Additionally, the ML techniques deal with correlated signal conditions better than subspace-based techniques.

Assume n observations that have $x_1, x_2, \ldots, x_n$ samples impending with unidentified probability density function $f_0(\cdot)$ from certain distribution. The joint density function of all observations can be defined as:

$$f(x_1, x_2, \ldots, x_n | \theta) = f(x_1 | \theta) \times f(x_2 | \theta) \times \cdots \times f(x_n | \theta) \qquad (3.12)$$

For a parametric function, Eq. (3.12) is called the likelihood, which is given by:

$$\mathcal{L}(\theta; x_1, \ldots, x_n) = f(x_1, x_2, \ldots, x_n | \theta) = \prod_{i=1}^{n} f(x_i | \theta) \qquad (3.13)$$

where the following log-likelihood function is employed:

$$\ln \mathcal{L}(\theta; x_1, \ldots, x_n) = \sum_{i=1}^{n} \ln f(x_i | \theta) \qquad (3.14)$$

The average log-likelihood estimator is then expressed as follows:

$$\hat{\ell} = \frac{1}{n} \ln \mathcal{L} \qquad (3.15)$$

In the model, this estimates the predictable log-likelihood of a particular observation.

In order to estimate the DOA using the ML estimator, the value of θ_0, which represents the required DOA is determined by finding the value of the angle that maximizes $\hat{\ell}(\theta; x)$, which describes the maximum likelihood estimator (MLE) of θ_0 as follows, if a maximum occurs:

$$\{\hat{\theta}_{\text{MLE}}\} \subseteq \{\arg\max_{\theta \in \Theta} \hat{\ell}(\theta; x_1, \ldots, x_n)\} \qquad (3.16)$$

This expression represent he estimated DOAE of the speakers in microphone array problems.

3.1.4 Local Polynomial Approximation Beamformer

Recently, an efficient beamforming technique using local polynomial approximation (LPA) has been developed and modified for different array geometries. Undeniably, this nonparametric LPA beamformer technique is first applied to DOA estimation by Katkovnik and Gershman [13], then it is generalized, modified and developed by Ashour et al. [2] and Elkamchouchi et al. [3].

It is worth noting that localizing and tracking multiple narrowband moving sources using a passive array are considered the fundamental problems in communication, sonar, radar, and microphone arrays. The conventional beamforming and high-resolution subspace techniques are established to exploit the benefits of temporal integration of array data for unmoving arrays and sources [14]. It assumes that only quite short series of observations are used for beamforming and estimation in non-stationary moving sources.

Conventional approaches fail and have deteriorated performance with the scenarios of moving sources. The LPA beamformer is quite different from the ML and the conventional beamformer [15]. Definitely, in the standard ML formulation, the source steering vectors are assumed to be time-invariant leading to different forms of the beamforming functions. The computational complexity of the LPA beamforming is M times higher than that of the conventional beamforming algorithm, where M is the number of points in the angular velocity domain. Typically, the LPA is a sliding window polynomial filtering (transform). The linear first degree LPA treats the discrete time 1D signal as sampled from an underlying continuous function within the selected window and uses loss function [16]. For DOA estimation of moving speech sources, the LPA beamformer estimates the time-varying DOA $\hat{\theta}(t)$ from a finite number N of the array observations $\mathbf{r}(t)$. For DOA estimation, let the speech source motion model within the observation interval using Taylor series [17]:

$$\theta(t + kT) = \theta(t) + \theta^{(1)}(t)(kT) + \frac{\theta^{(2)}(t)}{2}(kT)^2 + \frac{\theta^{(3)}(t)}{6}(kT)^3 + \cdots \tag{3.17}$$
$$= c_0 + c_1 kT + c_2(kT)^2 + c_3(kT)^3 + \cdots$$

where T is the sampling interval, and the parameters c_0 and c_1 will be used as estimates of the angle $\theta(t)$ and its derivative $\theta^{(1)}(t)$. The source trajectories are considered arbitrary functions of time that fit the nonparametric f piecewise continuous α-differentiable function, which is given by:

$$F_\alpha = \left\{ \left| \theta^{(\alpha)}(t) \right| \leq L_\alpha, \ \theta^{(\alpha)}(t) = \frac{d^\alpha \theta(t)}{dt^\alpha} \right. \tag{3.18}$$

For $\alpha = 0$, $|\theta(t)|$ is just restricted by the value L_0. For $\alpha = 1$ and 2, the velocity (first derivative) and the acceleration (second derivative) of $\theta(t)$ exist for almost every time instants and the absolute values of these derivatives have as upper bounds L_1 and L_2, respectively. The word "nonparametric" indicates that nothing is known about a parametric form of [18]. The source localization is ensured by a sliding weight-function (window) ω_h that discounts observations outside a neighborhood of the center t of the approximation.

Different kind of windows can be used, for example, the rectangular windows have equal weights for observations in the window. Nonrectangular windows, such as triangular, quadratic, usually prescribe higher weights to observations which are closer to the center t.

The window function can be expressed by:

$$\omega_h(kT) = (\frac{T}{h})\omega(\frac{kT}{h}) \tag{3.19}$$

where $\omega(v)$ is a real symmetric function $[\omega(v) = \omega(-v)]$ satisfying the following conventional properties:

$$\omega(v) \geq 0, \omega(0) = \max_v \omega(v), \int_{-\infty}^{\infty} \omega(v)dv = 1 \tag{3.20}$$

where the scaling parameter h determines the window length.

For linear 1D LPA beamformer, assume sufficiently short window, consequently, the third and later terms in Eq. (3.17) are insignificant, hence, the following model is obtained:

$$\theta(t + kT) = c_0 + c_1 kT \tag{3.21}$$

here $c_0 = \theta(t)$ and $c_1 = \theta^{(1)}(t)$ represent the instantaneous source DOA and angular velocity, respectively. So, the problem is to find the estimate $\hat{\mathbf{c}}$ of the vector $\mathbf{c} = (c_0, c_1)^T$ for each speech source of interest from a finite number of non-stationary array observations.

The loss function of the LPA using the weighted least squares approach in order to estimate the angle and its derivative is given by [6]:

$$G(t, \mathbf{c}) = \frac{1}{\sum_k \omega_h(kT)} \sum_k \omega_h(kT)\|\mathbf{r}(t + kT) - \mathbf{a}(\mathbf{c}, kT)s(t + kT)\|^2 \tag{3.22}$$

where $\mathbf{a}(\mathbf{c}, kT) = \mathbf{a}(c_0 + c_1 kT)$ and $\mathbf{e}(t + kT) = \mathbf{r}(t + kT) - \mathbf{a}(\mathbf{c}, kT)s(t + kT)$ is a residual of fitting the output $\mathbf{r}(t + kT)$ of the sensor by the corresponding output $\mathbf{a}(\mathbf{c}, kT)s(t + kT)$ of the steering vector and $\omega_h(kT)$ is the window.

The minimization of $G(t, \mathbf{c})$ with respect to the unknown deterministic waveform $s(t + kT)$ is expressed by:

$$\frac{\partial G}{\partial s^*(t + kT)} = \frac{-\omega_h(kT)}{\sum_k \omega_h(kT)}\mathbf{a}^H(\mathbf{c}, kT)\{\mathbf{r}(t + kT) - \mathbf{a}(\mathbf{c}, kT)s(t + kT)\} = 0 \tag{3.23}$$

Therefore, the estimate of the waveform $s(t + kT)$ is given by:

$$\hat{s}(t + kT) = \frac{\mathbf{a}^H(\mathbf{c}, kT)\mathbf{r}(t + kT)}{n} \tag{3.24}$$

where the property $\mathbf{a}^H(\mathbf{c}, kT)\mathbf{a}(\mathbf{c}, kT) = n$ is exploited. By substituting from Eq. (3.19) in Eq. (3.17), thus,

$$G(t, \mathbf{c}) = \frac{1}{\sum_k \omega_h(kT)} \sum_k \omega_h(kT) \left\{ \mathbf{r}^H(t + kT)\mathbf{r}(t + kT) - \frac{|\mathbf{a}^H(\mathbf{c}, kT)\mathbf{r}(t + kT)|^2}{n} \right\}$$

$$(3.25)$$

This function should be minimized over the vector parameter $\mathbf{c}$. Since, only the second term depends on $\mathbf{c}$, hence, the minimization of $G(t, \mathbf{c})$ is equivalent to the maximization of the LPA beamformer function, which is given by:

$$P(t, \mathbf{c}) = \frac{1}{n \sum_k \omega_h(kT)} \sum_k \omega_h(kT) |\mathbf{a}^H(\mathbf{c}, kT)\mathbf{r}(t + kT)|^2 \qquad (3.26)$$

This function is independent of the nature of $s(t)$, thus, if the transmitted signal is unknown, it will not affect this term in the algorithm. The maximization of $P(t, \mathbf{c})$ can be performed using the two-dimensional (2D) search over c_0 and c_1. Subsequently, this LPA beamformer can be protracted to multiple sources situations using direct superposition of particular responses to each source.

At the time instant t, the estimates of $\theta(t)$ and $\theta^{(1)}(t)$ as well as the value of the waveform $s(t)$ constitute a solution of the optimization problem:

$$(\hat{\theta}(t), \hat{\theta}^{(1)}(t), \hat{s}(t)) = \arg(\min_{\mathbf{c}, s(t)} G(t, \mathbf{c})) \qquad (3.27)$$

or

$$(\hat{\theta}(t), \hat{\theta}^{(1)}(t)) = \arg(\max_{\mathbf{c}} P(t, \mathbf{c})) \qquad (3.28)$$

From the preceding procedure, the DOA, and the velocity of the moving speaker can be estimated accurately using the LPA beamformer technique.

3.2 Optimization Algorithms in DOAE

In the near-field, numerous algorithms were implemented in the recent years for source localization. Several algorithms are based on subspace approaches, while, others use evolutionary computing methods. Practically, near-field case arises in innumerable situations, including microphone arrays for speech enhancement, seismic exploration, under water source localization, and ultrasonic imaging. For near-field source localization, several researchers proposed many approaches, such as the 2D MUSIC technique; high-order spectra (HOS) based algorithms; the weighted linear prediction technique; the ESPRIT technique; and the ML technique [19, 20]. However, in practical conditions, these proposed procedures are computationally heavy, some require extra computations for parameters pairing in case of multiple sources, in addition, closely spaced sources' localization suffer from poor estimation in low Signal to Noise Ratio (SNR).

Optimization algorithms and evolutionary computing methods, such Particle Swarm Optimization (PSO), Genetic algorithm (GA), Differential Evolution (DE), and Genetic programming (GP) proved their significance recently [21–27]. These methods achieve commanding global optimizers with avoiding being stuck in local minima. In addition, they can be hybridized to provide reliable and effective optimized solutions.

Using an array antenna for DOAE from the received signal is a critical topic in sonar, radar, communication systems, and microphone array systems. Traditional DOAE techniques including ML, MUSIC, root-MUSIC, ESPRIT have been used. Recently, the ML estimation using particle swarm optimization (PSO) [28], the genetic algorithm (GA)-based technique [29], and the evolutionary programming (EP)-based method [30] are developed. Choi [31] implemented a new DOAE scheme using PSO-based SPECC (Signal Parameter Extraction via Component Cancellation), where the optimization method supports the extraction process of the signal sources' amplitudes and incident angles that impinge on the sensor array. Sheikh et al. [32] employed differential evolution algorithm for range and DOAE of near-field narrow band sources that impose on a uniform linear array (ULA). During the optimization steps, the mean square error (MSE) is used as a fitness function. The results of DE are compared with the results of Genetic Algorithm (GA).

3.3 Time of Arrival Estimation Techniques

Speaker localization is concerned with locating the speaker position in a certain place according to the received sound signals from the MA. This process supports several real world applications including speech recognition, video conferencing [33], speech acquisition [34], hands-free voice communication [35], acoustic surveillance devices that require high quality of captured speech from the speakers [36]. Acoustical situation is degraded by background/additive noise, and distortion due to the speech signal reverberation from a speaker. In order to overwhelm such problem, the speech is recorded using microphones set, which require moving speaker's localization and tracking. Once the real speaker position is known, the MA can be electronically steered for high-quality speech acquisition. Moreover, the speaker localization is vital in the multi-speaker scenario. Time of arrival estimation (TDOA) localization scheme calculates the time delay estimation between each microphone's pair and the source using several techniques including the generalized cross correlation of maximum likelihood (GCC-ML), the generalized cross-correlation of phase transform (GCCPHAT), the Hilbert envelope of the LP residual and the linear prediction (LP) residual [37, 38].

The TDOA is considered the superior technique to compute the time delay estimation between each pair of microphones and the source. It is essential to acquire decent estimate of the time-delay even with corrupted signals by reverberation and noise. For TDE, the spectral features of speech signals are processed, which are affected by the speech degradations due to noise and reverberation [39, 40].

References

1. Ashour, A. S. (2005). *Smart antenna* (Doctoral dissertation, Ph.D. thesis). Faculty of Engineering, Tanta University, Egypt.
2. Ashour, A. S., Elkamchouchi, H. M., & Nasr, M. E. (2004, June). Planar array for accelerated sources tracking using local polynomial approximation beamformer. In *Antennas and Propagation Society International Symposium, 2004. IEEE* (Vol. 1, pp. 431–434). IEEE.
3. Elkamchouchi, H. M., Nasr, M. E., & Ashour, A. S. (2004, March). Modified LPA beamformer for localizing and tracking rapidly accelerated moving sources. In *Proceedings of the Twenty-First National Radio Science Conference, 2004. NRSC 2004* (pp. B4–1). IEEE.
4. Elkamchouchi, H. M., Nasr, M. E., & Ashour, A. S. (2004, March). Planar array for signal tracking using sliding window approach. In *Proceedings of the Twenty-First National Radio Science Conference, 2004. NRSC 2004* (pp. B10–1). IEEE.
5. Elkamchouchi, H. M., Nasr, M. E., & Ashour, A. S. (2005). Nonparametric approach for direction-of-arrival estimation of accelerated moving sources using cubic array. *Alexandria Engineering Journal, 44*(3), 401–411.
6. Ashour, A. S. (2014). LPA beamformer for tracking nonstationary accelerated near-field sources. *International Journal of Advanced Computer Science and Applications, 5*(3), 2–9.
7. Ashour, A. S., & Dey, N. (2016). Adaptive window bandwidth selection for direction of arrival estimation of uniform velocity moving targets based relative intersection confidence interval technique. *Ain Shams Engineering Journal*.
8. Brandstein, M., & Ward, D. (Eds.). (2013). *Microphone arrays: signal processing techniques and applications*. Berlin: Springer Science & Business Media.
9. Zeira, A., & Friedlander, B. (1996). Direction of arrival estimation using parametric signal models. *IEEE Transactions on Signal Processing, 44*(2), 339–350.
10. Stine, J. A. (1997). *Beamforming illustrations* (final report for the course project of EE381 k). Advanced Digital signal processing, pp. 1–9, July 1997.
11. Schell, S. V. (1994). Performance analysis of the cyclic MUSIC method of direction estimation for cyclostationary signals. *IEEE Transactions on Signal Processing, 42*(11), 3043–3050.
12. Ziskind, I., & Wax, M. (1988). Maximum likelihood location of multiple sources by alternating projection. *IEEE Transactions on Acoustics, Speech, and Signal Processing, 36,* 1553–1560.
13. Katkovnik, V., & Gershman, A. B. (2002). Performance study of the local polynomial approximation based beamforming in the presence of moving sources. *IEEE Transactions on Antennas and Propagation, 50*(8), 1151–1157.
14. Krim, H., & Viberg, M. (1996). Two decades of array signal processing research. *IEEE Signal Processing Magazine, 13*(4), 67–94.
15. Katkovnik, V. (2000). *Adaptive robust array signal processing for moving sources and impulse noise environment*, TICSP Series #11, December 2000.

16. Swindlehurst, A. L., & Kailath, T. (1992). A performance analysis of subspace-based methods in the presence of model errors—part I: the MUSIC algorithm. *IEEE Transactions on Signal Processing, SP-40,* 1758–1774.

17. Dietrich, C. B., Jr., Stutzman, W. L., Kim, B., & Dietze, K. (2000). Smart antennas in wireless communications: Base-station diversity and handset beamforming. *IEEE Antennas and Propagation Magazine, 42*(5), 142–151.

18. Hanssen, R. F. (2001). *Radar interferometry: Data interpretation and error analysis* (Vol. 2). Berlin: Springer Science & Business Media.

19. Grosicki, E., Abed-Meraim, K., & Hua, Y. (2005). A weighted linear prediction method for near-field source localization. *IEEE Transactions on Signal Processing, 53*(10), 3651–3660.

20. He, J., Ahmad, M. O., & Swamy, M. N. S. (2013). Near-field localization of partially polarized sources with a cross-dipole array. *IEEE Transactions on Aerospace and Electronic Systems, 49*(2), 857–870.

21. Poli, R., Kennedy, J., & Blackwell, T. (2007). Particle swarm optimization. *Swarm Intelligence, 1*(1), 33–57.

22. Rocca, P., Benedetti, M., Donelli, M., Franceschini, D., & Massa, A. (2009). Evolutionary optimization as applied to inverse scattering problems. *Inverse Problems, 25*(12), 123003.

23. Zhang, J., Zhan, Z. H., Lin, Y., Chen, N., Gong, Y. J., Zhong, J. H., … & Shi, Y. H. (2011). Evolutionary computation meets machine learning: A survey. *IEEE Computational Intelligence Magazine, 6*(4), 68–75.

24. Ashour, A. S., Samanta, S., Dey, N., Kausar, N., Abdessalemkaraa, W. B., & Hassanien, A. E. (2015). Computed tomography image enhancement using cuckoo search: a log transform based approach. *Journal of Signal and Information Processing, 6*(03), 244.

25. Dey, N., Ashour, A. S., Beagum, S., Pistola, D. S., Gospodinov, M., Gospodinova, E. P., et al. (2015). Parameter optimization for local polynomial approximation based intersection confidence interval filter using genetic algorithm: an application for brain MRI image de-noising. *Journal of Imaging, 1*(1), 60–84.

26. Saba, L., Dey, N., Ashour, A. S., Samanta, S., Nath, S. S., Chakraborty, S., … & Suri, J. S. (2016). Automated stratification of liver disease in ultrasound: An online accurate feature classification paradigm. *Computer Methods and Programs in Biomedicine, 130,* 118–134.

27. Chatterjee, S., Sarkar, S., Hore, S., Dey, N., Ashour, A. S., & Balas, V. E. (2017). Particle swarm optimization trained neural network for structural failure prediction of multistoried RC buildings. *Neural Computing and Applications, 28*(8), 2005–2016.

28. Jiankui, Z., Zishu, H., & Benyong, L. (2006). Maximum likelihood DOA estimation using particle swarm optimization algorithm. In *2006 CIE International Conference on Radar.*

29. Karamalis, P., Marousis, A., Kanatas, A., & Constantinou, P. (2001). Direction of arrival estimation using genetic algorithms. In *Vehicular Technology Conference, 2001. VTC 2001 Spring. IEEE VTS 53rd* (Vol. 1, pp. 162–166). IEEE.

30. Karamalis, P., Marousis, A., Kanatas, A., & Constantinou, P. (2001). Direction of arrival estimation using genetic algorithms. In *Vehicular Technology Conference, 2001. VTC 2001 Spring. IEEE VTS 53rd* (Vol. 1, pp. 162–166). IEEE.

31. Choi, I. S. (2014, January). Direction of Arrival Estimation Using Particle Swarm Optimization-based SPECC. In *Proceedings of the International Conference on Genetic and Evolutionary Methods (GEM)* (p. 1). The Steering Committee of the World Congress in Computer Science, Computer Engineering and Applied Computing (WorldComp).

32. Sheikh, Y. A., Ullah, R., & Zhonfu, Y. (2016). Range and direction of arrival estimation of near-field sources in sensor arrays using differential evolution algorithm. *International Journal of Computer Applications (IJCA), 139*(4), 16–20.

33. Zotkin, D., Duraiswami, R., Philomin, V., & Davis, L. S. (2000). Smart videoconferencing. In *2000 IEEE International Conference on Multimedia and Expo, 2000. ICME 2000* (Vol. 3, pp. 1597–1600). IEEE.

34. Nordholm, S., Claesson, I., & Grbić, N. (2001). Optimal and adaptive microphone arrays for speech input in automobiles. In *Microphone Arrays* (pp. 307–329). Berlin: Springer.

35. Jeannes, W. L. B., Scalart, P., Faucon, G., & Beaugeant, C. (2001). Combined noise and echo reduction in hands-free systems: A survey. *IEEE Transactions on Speech and Audio Processing, 9*(8), 808–820.
36. Raykar, V. C., Yegnanarayana, B., Prasanna, S. M., & Duraiswami, R. (2005). Speaker localization using excitation source information in speech. *IEEE Transactions on Speech and Audio Processing, 13*(5), 751–761.
37. DiBiase, J. H., Silverman, H. F., & Brandstein, M. S. (2001). Robust localization in reverberant rooms. In *Microphone Arrays* (pp. 157–180). Berlin: Springer.
38. Asaei, A., Taghizadeh, M., & Sameti, H. (2007). Speaker direction finding for practical systems: A comparison of different approaches. In *Third Annual IEEE BENELUX/DSP valley signal processing symposium* (No. EPFL-CONF-181697).
39. Liu, R., & Wang, Y. (2010). Azimuthal source localization using interaural coherence in a robotic dog: Modeling and application. *Robotica, 28*(7), 1013–1020.
40. Remaggi, L., Jackson, P. J., Coleman, P., Wang, W., Remaggi, L., Jackson, P. J., … & Wang, W. (2017). Acoustic reflector localization: Novel image source reversion and direct localization methods. *IEEE/ACM Transactions on Audio, Speech and Language Processing (TASLP), 25*(2), 296–309.

Chapter 4
Applied Examples and Applications of Localization and Tracking Problem of Multiple Speech Sources

In an enclosed environment, multi-source DOAE is an inspiring due to environmental noise, room reverberation, and the source spectra overlapping. The DOAE of acoustic sources is imperative for several applications including beamforming, automatic camera steering, robotics and surveillance. The incidence of background noise and reverberation should be considered in the real environment. Furthermore, DOAE of multiple and concurrent active sources is considered a challenging problem [1–4]. For these applications, conventional methods often exploit an omnidirectional microphone array with DOAE using phase-delay information between the microphones. Nevertheless, the conventional arrays necessitate a large aperture.

4.1 Simulation of LPA Beamformer

Assume a ULA of ten omnidirectional sensors with the half-wavelength inter-element spacing, and uncorrelated sources with SNR = 0 dB in a single sensor. For three signals $s_1(t)$, $s_2(t)$ and $s_3(t)$ be zero-mean white Gaussian with $\sigma = 1$, have the angles $\theta_i(t)$, i = 1, 2, 3. Assume rectangular window has N = 60 snapshots and sampling period of $T = 1$ s. Thus, the observation model is represented as follows [5–7]:

$$r(t + kT) = A(t + kT)s(t + kT) + e(t + kT) \tag{4.1}$$

$$A(t + kT) = [a(\theta_1(t + kT)), a(\theta_2(t + kT)), \ldots, a(\theta_q(t + kT))] \tag{4.2}$$

$$s(t) = [s_1(t), s_2(t), \ldots, s_q(t)]^T \tag{4.3}$$

© The Author(s) 2018
N. Dey and A. S. Ashour, *Direction of Arrival Estimation and Localization of Multi-Speech Sources*, SpringerBriefs in Speech Technology,
https://doi.org/10.1007/978-3-319-73059-2_4

Afterwards, the time-varying $\theta_i(t)$ should be estimated, where the time-argument is centered around the t of the LPA, where:

$$\theta_i(t + kT) = \theta_i(t) + \theta_i^{(1)}(t)kT \qquad (4.4)$$

Thus, $\theta_i(t + kT)$ is linear on time kT with the values of the direction $\theta_i(t)$ and the first derivative $\theta_i^{(1)}(t)$ for the time-instant t. The performance of the LPA beamformer versus the conventional beamformer is studied for different scenarios as follows for multi-speech signals from moving sources.

4.1.1   Case 1 (One Source Case)

Consider a source, moving with three different uniform velocities, namely $\theta^{(1)}(t) = (0, 1, 2)$ deg/sample, where $\theta(t) = 0°$. The LPA function is displayed in Figs. 4.1 through 4.4 as a 2D contour plot and 3D surface plot are illustrated with focusing on the location of interest. Figure 4.1 illustrates the LPA beamformer output of a single stationary source.

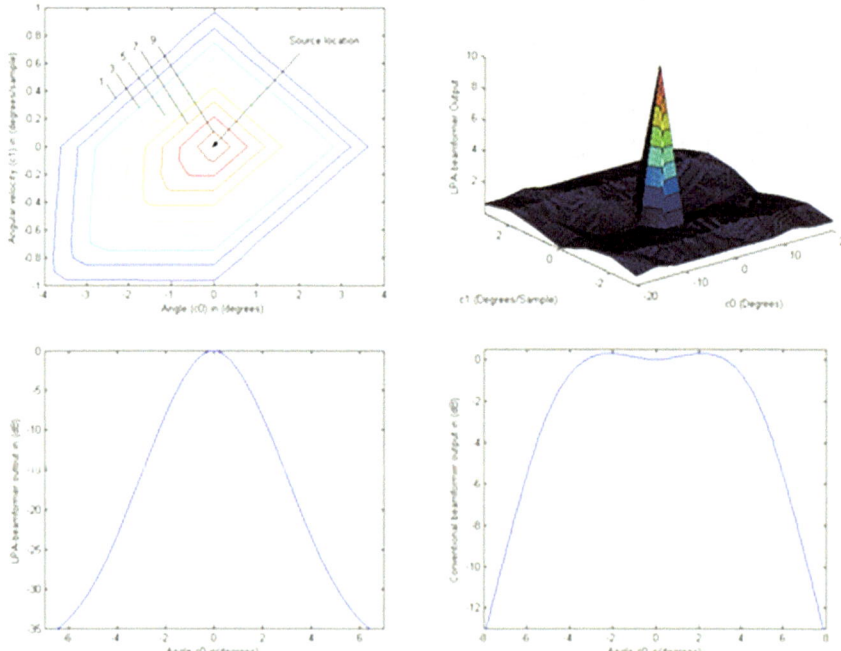

Fig. 4.1 A single stationary source case beamformers output at $\theta = 0$

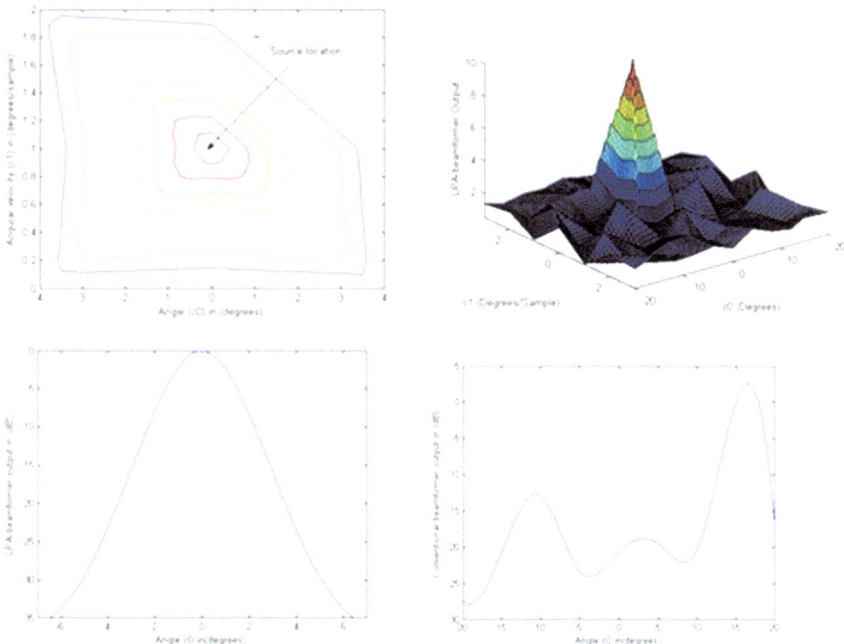

Fig. 4.2 Beamformers output for a single source case at $\theta = 0$ and $\theta^{(1)}(t) = 1$ deg/sample

Figure 4.1 shows the accurate source localization at $\theta = 0$ and $\theta^{(1)}(t) = 0$. Figure 4.2 illustrates the LPA beamformer output compared to the conventional beamformer P_{conv} for a single source case at $\theta = 0$ and $\theta^{(1)}(t) = 1$ deg/sample.

Figure 4.2 establishes that the LPA provided an accurate indication of the source location at $\theta = 0$. Also, P_{LPA} shows $\theta^{(1)}(t) = 1$, while, P_{conv} cannot indicate the source location, where the peak location is shifted. Figure 4.3 illustrates the comparative results of the LPA and the conventional beamformers for a single source case at $\theta = 0$ and $\theta^{(1)}(t) = 2$ deg/sample.

Figure 4.3 illustrates the accurate localization of the source using the LPA beamformer at $\theta = 0$ and $\theta^{(1)}(t) = 2$, while, P_{conv} cannot indicate the source location, where the output is degraded. Figure 4.4 demonstrates the beamformers output for a single source case at $\theta = 0$ and $\theta^{(1)}(t) = -1$ deg/sample.

Figure 4.4 establishes the achieved accurate localization of the single source that located at $\theta = 0$ and $\theta^{(1)}(t) = -1$ using the LPA beamformer, while, P_{conv} cannot indicate the source location. The peaks of all curves indicate the efficient localization of the source position using the LPA beamformer. The true, accurate values of $\theta(t)$ and $\theta^{(1)}(t)$, while the shape of the curves is slightly depending on the angular velocity. The LPA function P_{LPA} is expressed in term of c_0 i.e., θ as well as the conventional beamformer P_{conv}. The peaks of all the LPA curves show the true

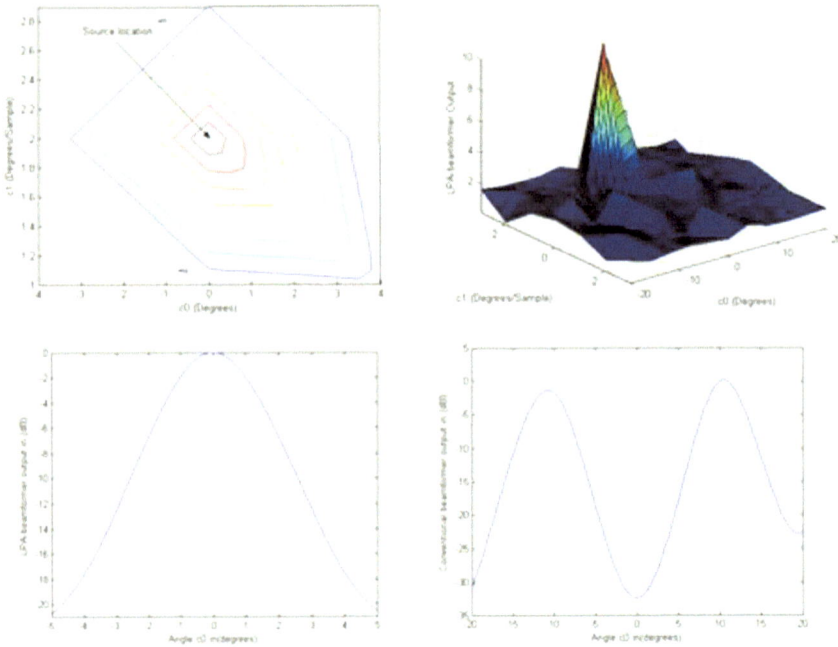

Fig. 4.3 Beamformers output for a single source case at $\theta = 0$ and $\theta^{(1)}(t) = 2$ deg/sample

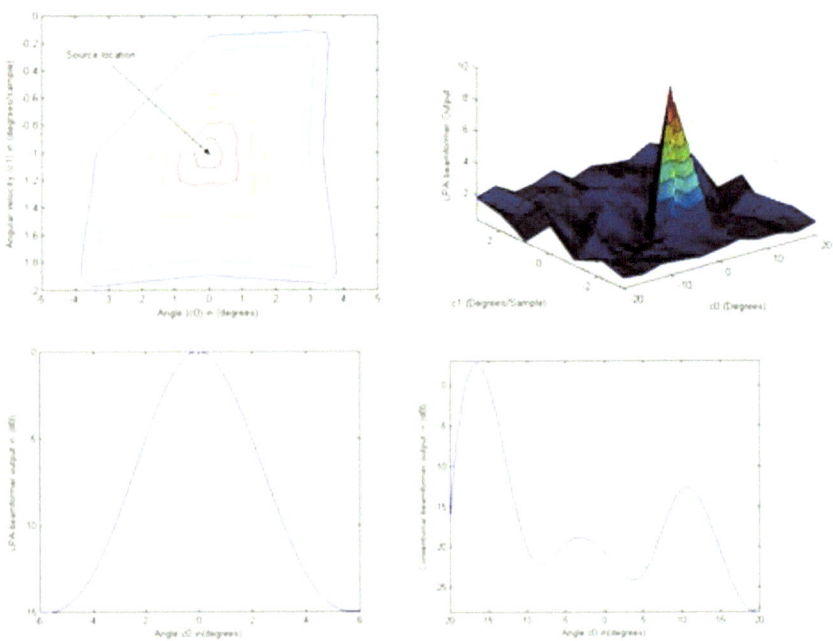

Fig. 4.4 Beamformers output for a single source case at $\theta = 0$ and $\theta^{(1)}(t) = -1$ deg/sample

value of $\theta(t)$ for all velocities. The increasing of the source velocity shifts the P_{conv} peaks from the accurate value of the angle, while a degradation of the one-peak of P_{conv} for larger values of $\theta^{(1)}(t)$ occurs in Fig. 4.3. Comparing the P_{conv} output in Figs. 4.1 and 4.4, it is clear that the peak is shifted to the right when the angular velocity is positive, while it is shifted to left when the angular velocity is negative. Moreover, the P_{LPA} has no side-lobes in a wide range around the peak, which indicates the source location, and its main lobe bandwidth is smaller than that of the conventional beamformer. This can be considered an advantage of the LPA beamformer compared to the conventional beamformer even for the case of unmoving sources. As the source moves with negative uniform angular velocity as in Fig. 4.4, the P_{LPA} beamformer indicates the source location correctly.

4.1.2   Case 2 (Well Separated Multi Sources Case)

Consider the case of three sources (microphones), where the resolution of the sources depends on the direction θ and can be given by:

Fig. 4.5 LPA beamformers output for a well separated sources at $\theta_1(t) = -16°$, $\theta_2(t) = 0, \theta_3(t) = 16°$ and they have the same angular velocity $\theta^{(1)}(t) = 1°/sample$

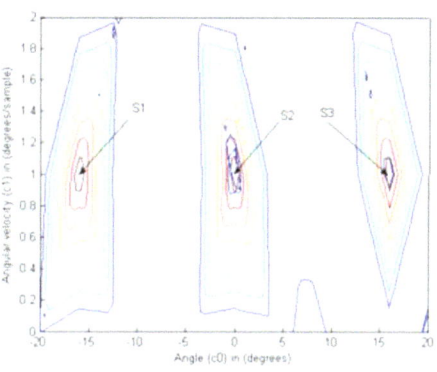

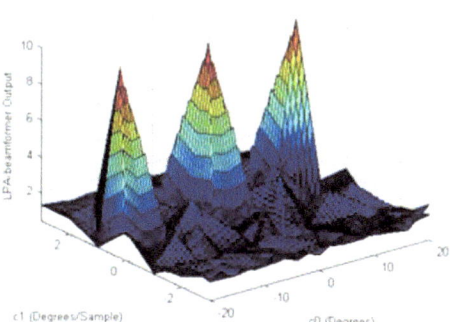

$$\Delta\theta \cong 1/|\Gamma \cos \theta|, \quad \Gamma = (n-1)\mathrm{d}/\lambda \qquad (4.5)$$

Assume well separated sources having $\theta_1(t) = -16°$, $\theta_2(t) = 0$, $\theta_3(t) = 16°$ and they have the same angular velocity $\theta^{(1)}(t) = 1°/sample$ as illustrated in Fig. 4.5.

Figure 4.5 establishes that the peaks of the LPA beamformer preserve their position at the actual location of the sources, which provide the right estimate of the source velocities. While the conventional beamformer is degraded and cannot localize the sources correctly where the angular velocity is considered (i.e., the sources are non-stationary).

4.2 Simulation of Frost Beamformers of Microphone Array

Sensor arrays are commonly employed for signal separation from noises based on the DOA information. Frost's beamformer has a significant role in speech processing for speaker localization. Each sensor in the Frost's beamformer array is followed by a transversal filter, which has weight as illustrated in Fig. 4.6 [8]. The beamformer output is the filter outputs' sum, where the weights updated by Frost's constrained least mean square (CLMS) procedure to minimize the mean square error of the output signal.

The whole Frost's beamformer system can be supplanted by one transversal FIR filter in the speech signal. The Matlab function in [9] is used to implement a time domain beamformers to recover speech signals by a microphone array of noisy microphone array measurements and to simulate an interference-dominant signal received. The Frost's beamformer is applied as it has superior interference suppression ability compared to the time-delay approach. The Frost beamformer

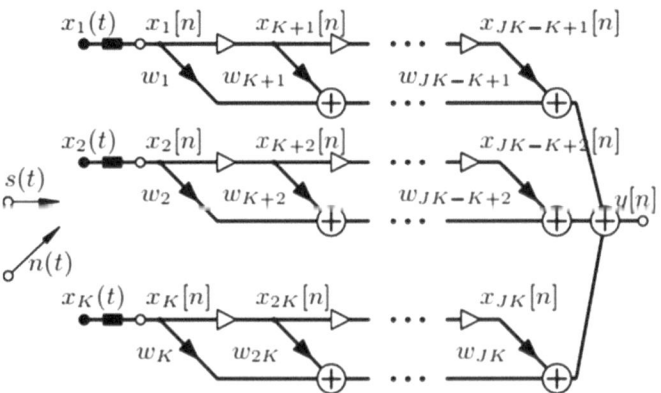

Fig. 4.6 Frost's beamformer configuration

robustness is achieved using diagonal loading. Several scenarios are applied to show the structure effect of the omnidirectional microphones configuration, the elements' spacing and the number of speech sources.

4.2.1   Case 1 (ULA of Ten Omnidirectional Microphones)

Assume a uniform linear array (ULA) 10 omnidirectional microphones to receive the speech signal. The array elements have 5 cm spacing, where multichannel signals are received by the MA. Two recorded speeches with one laughter recordings have been included, where the laughter audio refers to the interference. The azimuth and elevation directions of the speech signals are $(-30°, 0°)$ and $(-10°, 10°)$ of the first and second speech signals; respectively, while, the inter-ference (laughter) comes from the direction $(20°, 0°)$, which masks the speech signals. In addition, for each sensor, white noise signal of $1e^{-4}$ watts representing the thermal noise is considered. At the array origin, each input single-channel signal is received by a single microphone. Figure 4.7 represents the channel 3 received signal.

In order to compensate the arrival time differences (ATD) across the array, a time delay beamformer is applied with the coming signal from certain direction. A time delay (TD) beamformer is constructed to delineate a steering angle

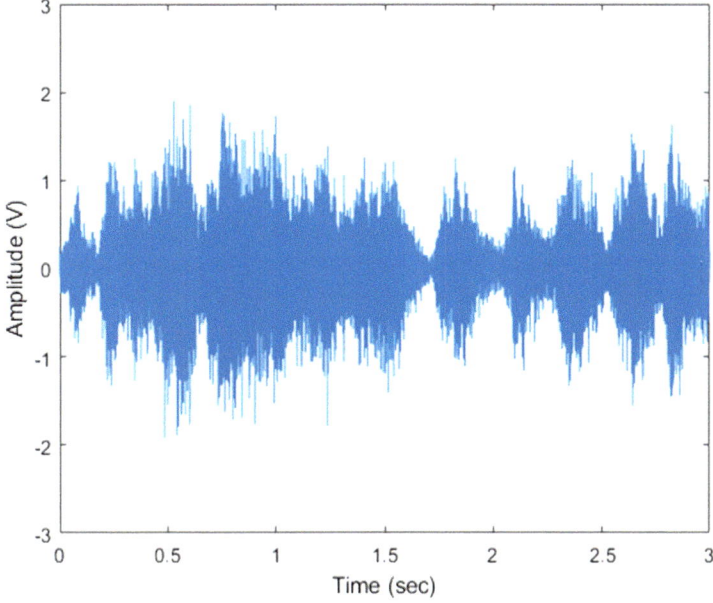

Fig. 4.7 The received speech signal at channel 3

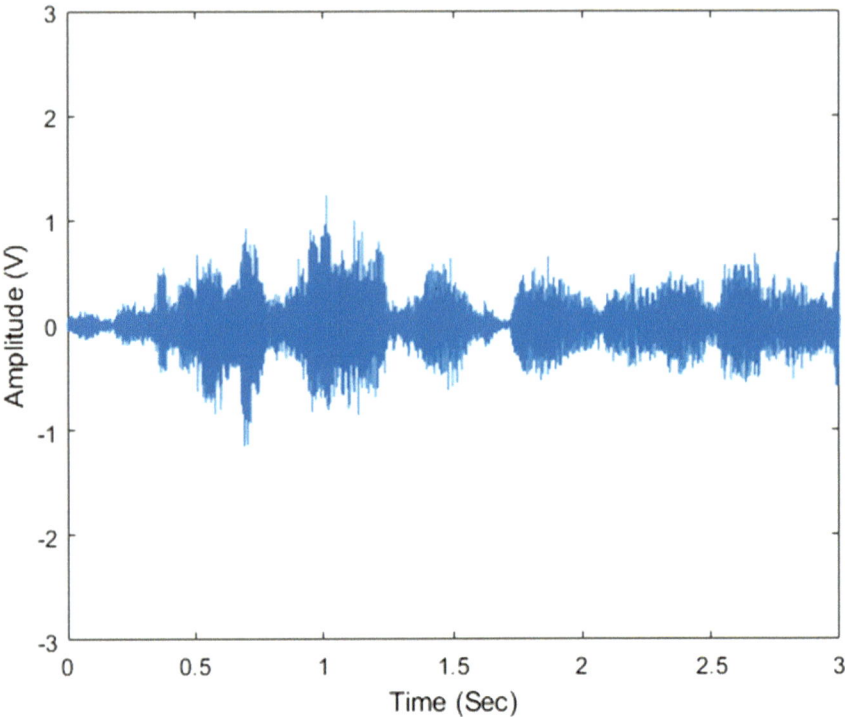

Fig. 4.8 The TD beamformer output

consistent to the first speech signal's incident direction. Figure 4.8 illustrated the TD beamformer output.

The speech enhancement can be reported by measuring the array gain representing the ratio of output to input signal-to-interference-plus-noise ratio (SINR). In addition, a Frost beamformer can be used to acquire superior beamformer performance, where the attached FIR filters to each sensor provided the Frost beamformer with more weights for suppressing the interference. Thus, nulls can be placed at the interference directions for superior interference suppression. The Frost beamformer achieved 14 dB array gain, which is 4.5 dB higher than that of the TD beamformer. Furthermore, the frost's beamformer can be used to steer the array in the direction of the second speech signal. Figure 4.9 illustrates the frost's beamformer output with diagonal loading to improve its performance.

In order to demonstrate the effect of the microphone array configuration and cumber of elements, the following scenarios are applied.

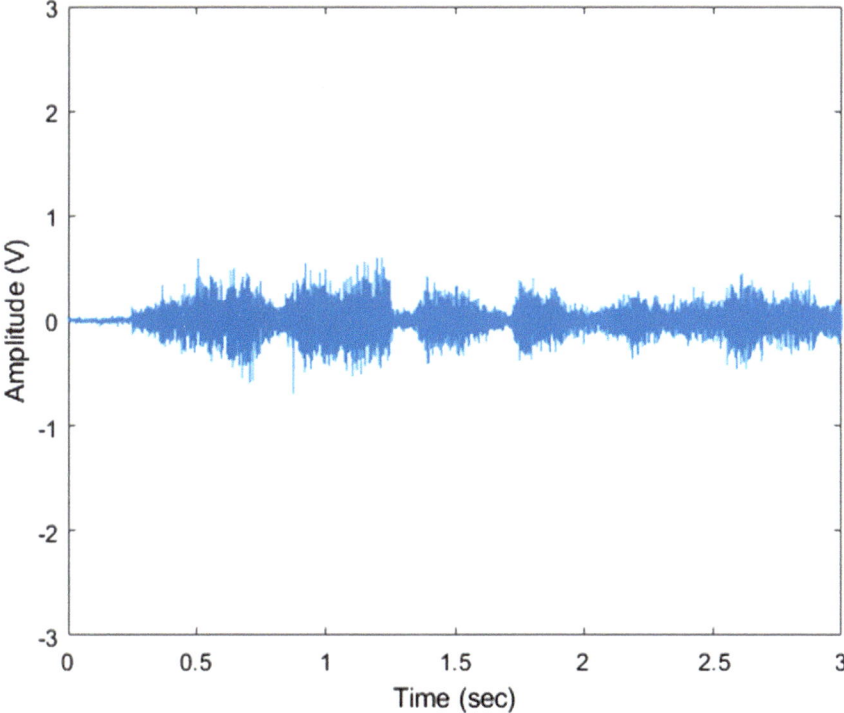

Fig. 4.9 The frost's beamformer output with diagonal loading

4.2.2   Case 2 (ULA of 5 Omnidirectional Microphones)

In this case, the same signals at the same directions are received using a ULA with five omnidirectional microphones to receive the speech signal is used instead of using ten elements. In addition, the spacing between the elements is increased to be 10 cm instead of 5 cm. Figures 4.10, 4.11 and 4.12 represent the channel 3 received signal, the TD beamformer output, and the frost's beamformer output; respectively.

In order to demonstrate the effect of the microphone array configuration, the following scenario is applied.

4.2.3   Case 2 (UCA of 5 Omnidirectional Microphones)

In this case, the same signals at the same directions are received using a uniform circular array (UCA) with five omnidirectional microphones to receive the speech

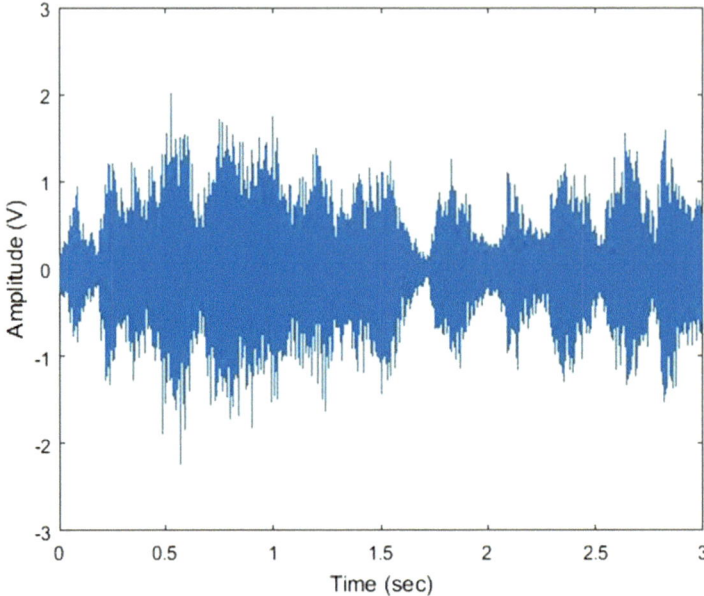

Fig. 4.10 The received speech signal at channel 3

Fig. 4.11 The TD
beamformer output

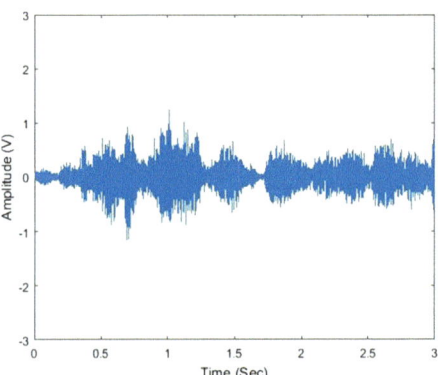

signal. In addition, the radius of the circular array is 1.5 cm. Figures 4.13, 4.14 and 4.15 represent the channel 3 received signal, the TD beamformer output, and the frost's beamformer output; respectively.

A comparative study between the TD beamformer and the frost's beamformer gain values from the preceding results of the different array configurations is illustrated in Fig. 4.16.

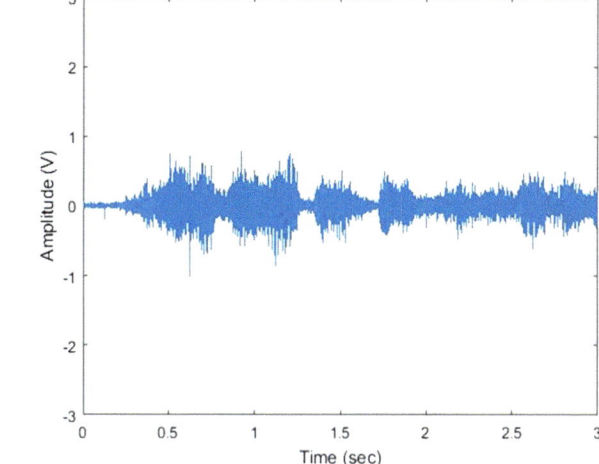

Fig. 4.12 The frost's beamformer output with diagonal loading

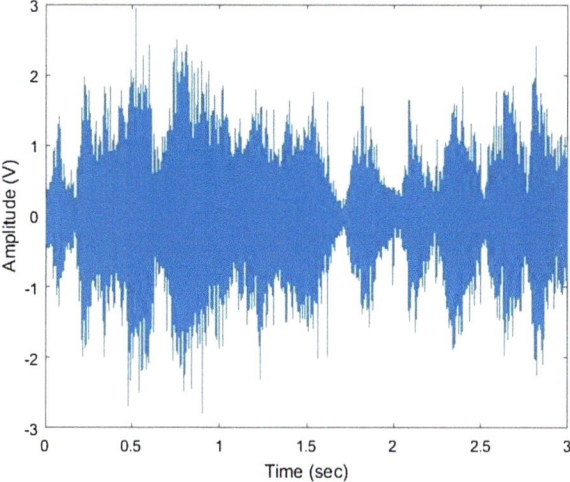

Fig. 4.13 The received speech signal at channel 3

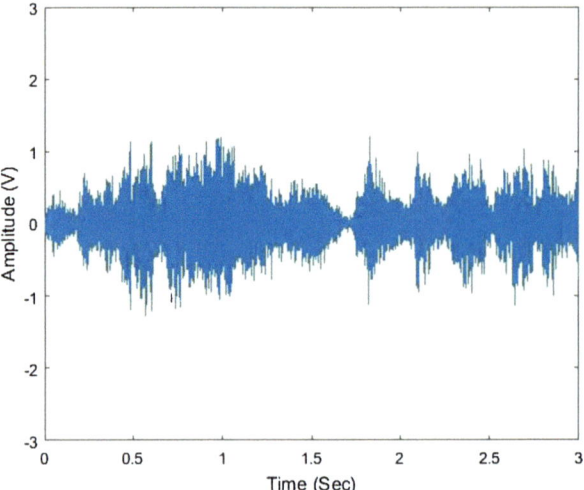

Fig. 4.14 The TD beamformer output

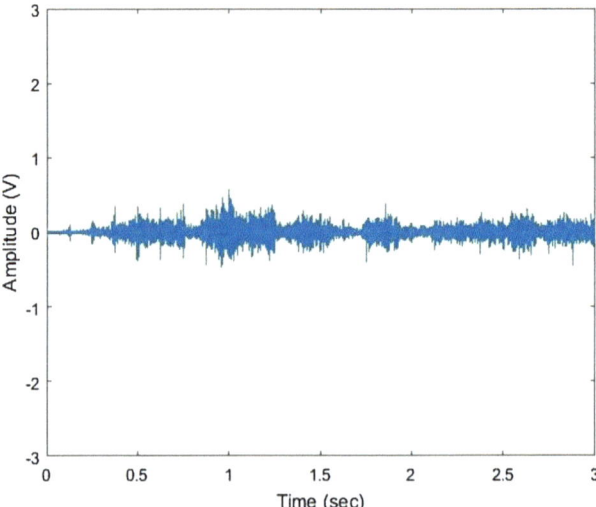

Fig. 4.15 The frost's beamformer output with diagonal loading

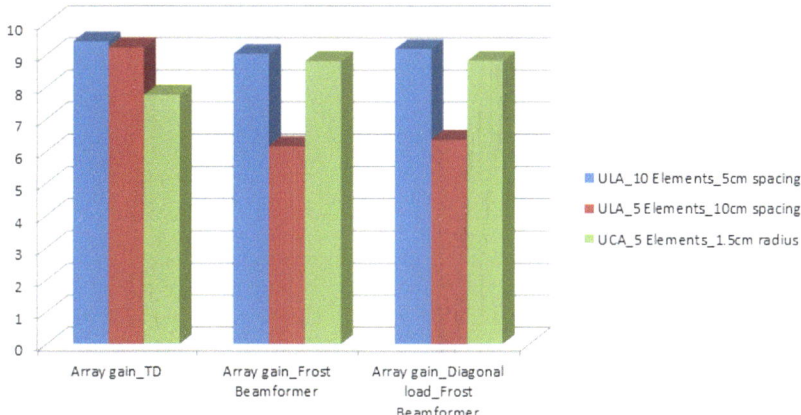

Fig. 4.16 The array gain using different configurations

4.3 Linear Microphone Array for Live Direction of Arrival Estimation

The 4 built-in microphones of the Microsoft Kinect™ are recognized using Matlab 2017 to estimate the linear coordinates [10]. The applied procedure is used independently with pairs of microphones to estimate the DOA, which are then combined to determine a single live DOA output. As the inter-microphone distance increases, DOAE sensitivity increases correspondingly. A bespoke arrow-based polar visualization (Fig. 4.17) is used to illustrate the DOA estimation of the sound source using multiple microphone pairs within the linear array, where the four microphone positions at [−0.088, 0.042, 0.078, 0.11].

Fig. 4.17 The polar plot of the DOAE

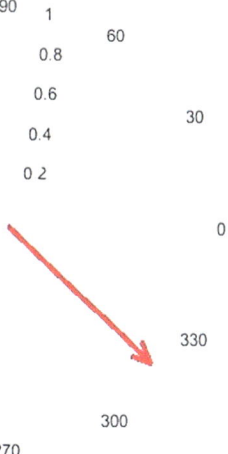

Fig. 4.18 The polar plot of
the DOAE

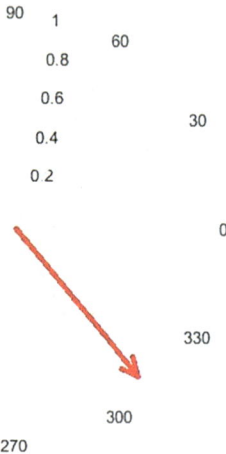

Figure 4.18 illustrates the polar plot of the DOAE when changing the microphone positions to be at [−0.05, 0.062, 0.098, 0.19].

References

1. Di Claudio, E. D., Parisi, R., & Orlandi, G. (2000). Multi-source localization in reverberant environments by ROOT-MUSIC and clustering. In *2000 IEEE International Conference on Acoustics, Speech, and Signal Processing, 2000. ICASSP'00. Proceedings* (Vol. 2, pp. II921–II924). IEEE.
2. Talantzis, F., Constantinides, A. G., & Polymenakos, L. C. (2005). Estimation of direction of arrival using information theory. *IEEE Signal Processing Letters, 12*(8), 561–564.
3. Zhong, X., & Premkumar, A. B. (2012). Particle filtering approaches for multiple acoustic source detection and 2-D direction of arrival estimation using a single acoustic vector sensor. *IEEE Transactions on Signal Processing, 60*(9), 4719–4733.
4. Wu, K., Reju, V. G., & Khong, A. W. (2015, April). Multi-source direction-of-arrival estimation in a reverberant environment using single acoustic vector sensor. In *2015 IEEE International Conference on Acoustics, Speech and Signal Processing (ICASSP)* (pp. 444–448). IEEE.
5. Ashour, A. S. (2005). *Smart antenna* (Doctoral dissertation, Ph.D. thesis). Faculty of Engineering, Tanta University, Egypt.
6. Ashour, A. S., Elkamchouchi, H. M., & Nasr, M. E. (2004, June). Planar array for accelerated sources tracking using local polynomial approximation beamformer. In *Antennas and Propagation Society International Symposium, 2004. IEEE* (Vol. 1, pp. 431–434). IEEE.
7. Ashour, A. S. (2014). LPA beamformer for tracking nonstationary accelerated near-field sources. *International Journal of Advanced Computer Science and Applications, 5*(3), 2–9.
8. Sovka, P., & Strupl, M. (2003). Analysis and simulation of frost's beamformer. *Radio engineering, 12*(2), 1–9.
9. https://www.mathworks.com/help/phased/examples/acoustic-beamforming-using-a-microphone-array.html.
10. https://www.mathworks.com/help/audio/examples/live-direction-of-arrival-estimation-with-a-linear-microphone-array.html.

Chapter 5
Challenges and Future Perspectives in Speech-Sources Direction of Arrival Estimation and Localization

DOAE is essential in several applications including automatic steering of video camera and multiparty teleconferencing for steering and beamforming in order to suppress reverberation and noise as well as to improve the speech intelligibility [1–7]. It has numerous challenges and future perspectives in speech-sources DOAE and localization. Acoustic source signal generally suffers from multiple reflections and ambient noise that significantly degrade the localization performance of the TDOA techniques with sound source using only two microphones.

In speech based systems, smart antenna systems have an imperative role after determining the speech signal's DOA. The smart antenna is considered to generate a beam toward the speaker and null toward the noise and the interferer. In addition, it enhances the capacity as well as the security [8]. Thus, in heavily faded areas, it becomes an active research area to determine the weaker coherent/non-coherent signals with low Signal-to-Ratio (SNR). Moreover, for multiple speech sources, microphone DOAE methods should be developed to exploit speech precise properties, such as the sparsity in time-frequency spectrum [9]. Using two-microphone array for sound source localization is a challenging topic with extensive potential in mobile devices, video conferencing, and robotics. A probability distribution of the source's location is considered according to the observed time-differences of arrival between the sound signals in order to estimate the actual source positions. Nevertheless, these procedures assume a specified number of sound sources [10].

One of the blooming domains is the use of DOAE and sound source localization (SSL) in robotics to associate the survival behaviors during the interaction between humans and robots [11]. Moreover, conventional DOAE techniques still cannot achieve quite dependably performance in low SNR conditions. In noisy environments, increasing the robustness of the DOA estimator for human speech becomes challenging. Xue and Liu [12] proposed a sub-band weighting procedure, where for each channel; the speech signal is passed over a Gammatone filter-bank to attain a set of time-domain sub-band signals. Afterwards, based on a new cost function, the TDOA estimation is determined in each sub-band. In order to emphasize the estimation results, the sub-band weight is calculated with high probability containing

© The Author(s) 2018
N. Dey and A. S. Ashour, *Direction of Arrival Estimation and Localization of Multi-Speech Sources*, SpringerBriefs in Speech Technology, https://doi.org/10.1007/978-3-319-73059-2_5

speech signals. Lastly, the DOA is determined using the estimated TDOA and according to the microphone array geometry.

Mainly, the efficiency of speakers' localization and DOAE is based on the advancement in the smart antenna technology that is based on digital signal processing procedures. Smart antenna systems are able to locate and to track signals from the speakers as well as the interferers for dynamically adjusts the antenna pattern for reception enhancement of the signal-of-interest direction, while minimizing interference in the signal-of-not-interest direction. Thus, the smart antenna system performance depends on the competence of digital signal processing procedures. Numerous DOAE procedures are employed in adaptive array smart antenna to localize the desired signal. On the antenna array, the number of incident plane waves is estimated by the DOA techniques as well as their angle of incidence [13]. Several DOA techniques can be employed, such as the ML, MUSIC, Root-MUSIC, ESPRIT, and LPA. Recently, the global optimization procedures, including evolutionary strategies (ES), genetic algorithms (GA), particle swarm optimization (PSO), and evolutionary programming (EP) have emerged as robust and effective in several applications in different domains [14–23]. Consequently, employing such optimization algorithms to enhance the DOA estimation techniques for speakers' localization using different microphone array configurations can be considered a future perspective mainly in the noisy environment. In a sensor network, distributed MAs offer an original way to find snipers [24]. This processing type provides the opportunity for novel and enhanced applications.

References

1. Boulmaiz, A., Messadeg, D., Doghmane, N., & Taleb-Ahmed, A. (2017). Design and implementation of a robust acoustic recognition system for waterbird species using TMS320C6713 DSK. *International Journal of Ambient Computing and Intelligence (IJACI), 8*(1), 98–118.
2. Bureš, V., Tučník, P., Mikulecký, P., Mls, K., & Blecha, P. (2016). Application of ambient intelligence in educational institutions: visions and architectures. *International Journal of Ambient Computing and Intelligence (IJACI), 7*(1), 94–120.
3. Suwais, K. (2017). Assessing the utilization of automata in representing players' behaviors in game theory. *Game Theory: Breakthroughs in Research and Practice: Breakthroughs in Research and Practice, 106*.
4. Alenljung, B., Lindblom, J., Andreasson, R., & Ziemke, T. (2017). User experience in social human-robot interaction. *International Journal of Ambient Computing and Intelligence (IJACI), 8*(2), 12–31.
5. Juneja, D., Singh, A., Singh, R., & Mukherjee, S. (2017). A thorough insight into theoretical and practical developments in multiagent systems. *International Journal of Ambient Computing and Intelligence (IJACI), 8*(1), 23–49.
6. Trabelsi, I., & Bouhlel, M. S. (2016). Comparison of several acoustic modeling techniques for speech emotion recognition. *International Journal of Synthetic Emotions (IJSE), 7*(1), 58–68.

7. Trabelsi, I., & Bouhlel, M. S. (2017). Feature selection for GUMI kernel-based SVM in speech emotion recognition. In *Artificial intelligence: concepts, methodologies, tools, and applications* (pp. 941–953). IGI Global.

8. Dhar, A., Senapati, A., & Roy, J. S. (2016). Direction of arrival estimation for smart antenna using a combined blind source separation and multiple signal classification algorithm. *Indian Journal of Science and Technology, 9*(18).

9. Zhang, W., & Rao, B. D. (2009, April). Two microphone based direction of arrival estimation for multiple speech sources using spectral properties of speech. In *IEEE International Conference on Acoustics, Speech and Signal Processing, 2009. ICASSP 2009* (pp. 2193–2196). IEEE.

10. Escolano, J., Xiang, N., Perez-Lorenzo, J. M., Cobos, M., & Lopez, J. J. (2014). A Bayesian direction-of-arrival model for an undetermined number of sources using a two-microphone array. *The Journal of the Acoustical Society of America, 135*(2), 742–753.

11. Meza, I., Rascon, C., Fuentes, G., & Pineda, L. A. (2016). On indexicality, direction of arrival of sound sources, and human-robot interaction. *Journal of Robotics, 2016.*

12. Xue, W., & Liu, W. (2012). Direction of arrival estimation based on subband weighting for noisy conditions. In *Thirteenth Annual Conference of the International Speech Communication Association.*

13. Dongarsane, C. R., & Jadhav, A. N. (2011). Simulation study on DOA estimation using MUSIC algorithm. *International Journal of Technology And Engineering System (IJTES), 2* (1), 54–57.

14. Kausar, N., Palaniappan, S., Samir, B. B., Abdullah, A., & Dey, N. (2016). Systematic analysis of applied data mining based optimization algorithms in clinical attribute extraction and classification for diagnosis of cardiac patients. In *Applications of intelligent optimization in biology and medicine* (pp. 217–231). Berlin: Springer International Publishing.

15. Jagatheesan, K., Anand, B., Baskaran, K., Dey, N., Ashour, A. S., & Balas, V. E. (2018). Effect of nonlinearity and boiler dynamics in automatic generation control of multi-area thermal power system with proportional-integral-derivative and ant colony optimization technique. In *Recent advances in nonlinear dynamics and synchronization* (pp. 89–110). Cham: Springer.

16. Samanta, S., Choudhury, A., Dey, N., Ashour, A. S., & Balas, V. E. (2016). Quantum inspired evolutionary algorithm for scaling factors optimization during manifold medical information embedding. *Quantum inspired computational intelligence: Research and applications.* North Holland, NY: Elsevier.

17. Naik, A., Satapathy, S. C., Ashour, A. S., & Dey, N. (2016). Social group optimization for global optimization of multimodal functions and data clustering problems. *Neural Computing and Applications*, 1–17.

18. Beagum, S., Dey, N., Ashour, A. S., Sifaki-Pistolla, D., & Balas, V. E. (2017). Nonparametric de-noising filter optimization using structure-based microscopic image classification. *Microscopy Research and Technique, 80*(4), 419–429.

19. Jagatheesan, K., Anand, B., Dey, K. N., Ashour, A. S., & Satapathy, S. C. (2017). Performance evaluation of objective functions in automatic generation control of thermal power system using ant colony optimization technique-designed proportional–integral–derivative controller. *Electrical Engineering*, 1–17.

20. Chakraborty, S., Dey, N., Samanta, S., Ashour, A. S., Barna, C., & Balas, M. M. (2017). Optimization of non-rigid Demons registration using cuckoo search algorithm. *Cognitive Computation*, 817–826.

21. Kaliannan, J., Baskaran, A., Dey, N., & Ashour, A. S. (2016). Ant colony optimization algorithm based PID controller for LFC of single area power system with non-linearity and boiler dynamics. *World Journal of Modelling and Simulation, 12*(1), 3–14.

22. Dey, N., Samanta, S., Chakraborty, S., Das, A., Chaudhuri, S. S., & Suri, J. S. (2014). Firefly algorithm for optimization of scaling factors during embedding of manifold medical information: An application in ophthalmology imaging. *Journal of Medical Imaging and Health Informatics, 4*(3), 384–394.
23. Dey, N., Ashour, A. S., Shi, F., & Sherratt, R. S. (2017). Wireless capsule gastrointestinal endoscopy: Direction of arrival estimation based localization survey. *IEEE Reviews in Biomedical Engineering.*
24. Lindgren, D., Wilsson, O., Gustafsson, F., & Habberstad, H. (2009, July). Shooter localization in wireless sensor networks. In *12th International Conference on Information Fusion, 2009. FUSION'09* (pp. 404–411). IEEE.

Chapter 6
Conclusion

Estimating the DOA of speakers and speech signals are a wide-ranging problem of interest in acoustic signal processing. From the different literatures, it is concluded that beamforming based on optimization for DOA estimation is advantageous. Significant knowledge about the MA is essential before developing the DOAE techniques. Copious applications recommended the use of large arrays in the order of greater than 100 elements in auditoriums, while small arrays of 2 or 3 elements are recommended for mobile telephones and hearing aids. Besides, the microphone array technology is extensively realistic in surveillance, and speech recognition. Conventional techniques have been employed for MAs include fixed spatial filters, such as optimal beamformer, adaptive beamformer, and frequency invariant beam-formers. Such array methods assume either calibration signal knowledge or model knowledge in addition to localization information for their design. Accordingly, they typically embrace some form of localization and tracking along with the beam-forming methods. Currently, contemporary techniques, exhausting time, frequency masking and blind signal separation (BSS) techniques have enticed the researchers' attention. These methods are less reliant on localization and array model as well as the speech signals' statistical properties, including the non-stationarity, sparseness, and non-Gaussianity.

From the theoretical perspective, the spatial diversity is considered the foremost advantage of the multiple microphones, which is an efficient tool to combat reverberation, noise, and interference. In the speech signal (target), the used sus-taining physical feature is the difference in the coherence versus the noise field and for understanding the striving in the enhancement of the highly reverberant speech of the received microphone signals.

Traditional techniques such as hand-free operation of MAs, frequency invariant beamforming, and source localization are developed for efficient DOAE. Small size microphone arrays have numerous applications for hearing aids, close up micro-phones, and mobile terminals. The novelty in representing small size arrays sup-ports the suppression of multiple interferers. Abnormalities in speech stemming and noise from processing are principally unavoidable.

© The Author(s) 2018
N. Dey and A. S. Ashour, *Direction of Arrival Estimation and Localization of Multi-Speech Sources*, SpringerBriefs in Speech Technology, https://doi.org/10.1007/978-3-319-73059-2_6